EIGHTEENTH CENTURY EQUITATION

# MILITARY EQUITATION:
## OR,
# A METHOD OF BREAKING HORSES,
# AND TEACHING SOLDIERS TO RIDE

*Henry Herbert, 10th Earl of Pembroke*

&

# A TREATISE
# ON MILITARY EQUITATION

*William Tyndale*

Edited, with Introduction and Notes, by
*Charles Caramello*

Available at www.XenophonPress.com.

Copyright © 2018 by Xenophon Press LLC. All rights reserved. No part of this work may be reproduced or transmitted in any form or by any means, electronic or mechanical, including photocopying, or by any information storage or retrieval system except by a written permission from the publisher.

Front cover image from Tyndale's *A Treatise on Military Equitation*, plate page 41.
Back cover image from Pembrokes *Military Equitation: or, A Method of Breaking Horses, and Teaching Soldiers to Ride,* plate 4

Published by Xenophon Press LLC
7518 Bayside Road, Franktown, Virginia 23354-2106, U.S.A.
xenophonpress@gmail.com

ePub edition:
ISBN-13: 978-1948717-04-5

Print Edition:
ISBN-13: 978-1948717-03-8

# CONTENTS

Publisher's Foreword ............................................................................. i

Introduction ............................................................................................ iii

Acknowledgements ............................................................................... ix

Bibliographical Notes ........................................................................... xi

## MILITARY EQUITATION: OR, A METHOD OF BREAKING HORSES, AND TEACHING SOLDIERS TO RIDE

    Explanatory Notes

## A TREATISE ON MILITARY EQUITATION

    Explanatory Notes

Xenophon Press Library

# PUBLISHER'S FOREWORD

The benefit of these two historical texts cannot be overstated. The very fact that they are military manuals makes them distillations of the more elaborate descriptions found in de la Guérinière (*Ecole de Cavalerie The Complete Part II*, 2015), Newcastle, and the like. I often say: "riding is simple, but not easy."

Pembroke advocated for a better system of horsemanship in both military and non-military venues. His work was written to address shortfalls that he perceived were rampant in horsemanship.

In any epoch, including our own, trends of both malpractice and best practices co-exist simultaneously. Those who practice well seek out education, constantly improve, learn from their mistakes and the past, and seek mentors in horsemanship. As soon as horsemanship becomes dogmatic, and we think we know everything that there is to learn and stop seeking complementary knowledge, the creativity dies. Ours is a living science and art informed by new discoveries as well as founded on tried and tested principles of the past.

Tyndale further simplifies Pembroke much the same way General de Lagarenne simplified Faverot de Kerbrech's summary of Francois Baucher. The tradition of summarizing and recapitulating the ideas of others deepens our understanding. Our re-reading of these texts verifies our own insights and experiences in the living art today.

Charles Caramello has added brilliant commentary to assist understanding of historical context and he provides necessary definitions to unfamiliar terms. Dr. Caramello's introduction, notes, and citations make these two works very approachable.

Grounding our current beliefs and practices in historical precedent gives trainers and teachers confidence and students reassurance that they are adhering to sound principles. We should not practice simply because someone told us to do so. We should practice because we understand horsemanship. This understanding is enriched through repetition and hearing similar and new concepts expressed in different ways.

The classical rider is the educated rider. This is our *raison d'être* at Xenophon Press. We are dedicated to preserving and presenting valuable media both past and present that enrich the living horsemanship of our audience. We will continue to bring important works to traditional print as well as digital and video.

*Richard F. Williams*
*Publisher*
*Xenophon Press*

# INTRODUCTION

## CONTEXT

The growing number of books on horsemanship published in Europe in the 16th and 17th centuries gained momentum in the 18th century (and reached peak velocity in the 19th century). Scores of works were published in Great Britain alone between 1700 and 1800—some of them English translations of continental writing, most of them English language in origin. Each work fell somewhere on a grid defined by *horsemanship, farriery, dressage,* and *equitation*. Horsemanship, generally speaking, encompassed the latter two, and farriery both the shoeing and medical care of horses.

The large number of works that treated equine breeding and care speak to the ubiquity of the horse in 18th century agriculture, industry, transportation, sport, and warfare. Reflecting the rationalistic, scientific, and encyclopedic *zeitgeist*, works in those fields claimed reasoned thought and empirical evidence as their bases, and comprehensive "useful knowledge" as their value. Examples include Jacques de Solleysel, *The Compleat Horseman: or, Perfect Farrier* (1664, translated 1696, revised 1702) and Gervase Markham, *Markham's Masterpiece: Containing All Knowledge Belonging to the Smith, Farrier, or Horse-leach* (1703); John Reeves, farrier, *The Art of Farriery in Both Theory and Practice* (1758), and Thomas Wallis, surgeon, *The Farrier's and Horseman's Complete Dictionary* (1775).

Comparable in number, treatises on dressage and equitation also claimed experiential validity and utilitarian value, but with a pronounced accent on aesthetics—as suggested by the recurring terms in titles such as J.L. Jackson, *The Art of Riding; or Horsemanship Made Easy* (1765); Charles Hughes, *The Compleat Horseman; or, the Art of Riding Made Easy* (1772); Laurence O'Reilly, *The Art of Horsemanship* (1780); and John Adams, *Analysis of Horsemanship, Teaching the Whole Art of Riding* (1799 and 1805). Clustered in the second half of the century, these tracts all owe debts to William Cavendish's seminal work, *La méthode nouvelle et Invention extraordinaire de dresser les chevaux* (1658), republished in French in 1737 and translated into English in 1743, and to his companion work, *A New Method, and Extraordinary Invention, to Dress Horses* (1667).

Mounted warfare, of course, played a central and critical role in European affairs of state from the 16th through the 19th centuries, and essentially all English works on horsemanship throughout that period—beginning with Thomas Blundeville's *The Arte of Ryding and Breakinge Greate Horses* (1660)—applied either directly or indirectly to military usage. Salient among these works were a stream of studies focused on the British cavalry, particularly the light cavalry, or "light horse," that emerged in the 17th century and evolved over the course of the 18th century, firmly cementing its value by mid-century.

These works cluster late in the century, on the eve of "the Napoleonic period, 1796-1815, [that] marked a revolution in the politics of Europe and in the history of warfare," a moment when "the cavalry arm reached the highest point of its popular and professional acclaim" (DiMarco, 193-94). They include both official publications and unofficial treatises. The latter range from broad works *on* the discipline, such as Robert Hinde's *The Discipline of the Light-Horse* (1778) and Captain L. Neville's *A Treatise on the Discipline of Light Cavalry* (1796), to more focused works on equitation *for* the discipline, such as the Earl of Pembroke's *A Method of Breaking Horses, and Teaching Soldiers to Ride* (1761 and later editions) and William Tyndale's *A Treatise on Military Equitation* (1797).

## PEMBROKE'S *MILITARY EQUITATION*

Henry Herbert, tenth Earl of Pembroke (1734-94), self-described as "horse mad" since youth, attended riding academies abroad, and entered British cavalry service in 1752. Rising apace through the ranks in the King's Dragoon Guards and 1st Foot Guards, Pembroke was appointed lieutenant-colonel in the 15th Light Dragoons in 1759. Joining his regiment in 1760 in Germany, during the Seven Years' War, he left regimental service, in the same year, to command a cavalry brigade. He was appointed to the staff as major-general in 1761, and promoted lieutenant-general in 1770 and general in 1782. Pembroke concurrently held appointment as colonel of the 1st (Royal) Dragoons from 1764 until his death in 1794.

A wealthy and raffish aristocrat who was "devoted to horses [and] dogs, enjoyed travelling and shooting, [and] had some interest in music and the arts" (Screen, *ODNB*), Pembroke also wrote the influential treatise, *A Method of Breaking Horses, and Teaching Soldiers to Ride* (1761 and 1762), revised and retitled, *Military Equitation; or, A Method of Breaking Horses, and Teaching Soldiers to Ride* (1778 and 1793). It addressed, as Pembroke states in his dedication to the King, "the wretched system of Horsemanship, that at present prevails in the ARMY," a system that leaves men and horses unprepared (and thus, implicitly, the King's interests ill served) "for want of proper instructions and intelligence in this Art."

Written for "the use of the Cavalry," and representing only the "outlines" of a longer work in progress, *Military Equitation* offers a program, with specific lessons, for the proper training of military horse and rider. It comprises, in short, theory and application, principles and practices, with the emphasis on *theory* and *principles*. Its goals, and its audiences, appear to be twofold: as a work of advocacy for the resources and means to improve programs in horsemanship, it speaks to a spectrum of politicians and, especially, staff officers; more important, as a work of instruction designed to improve the *quality* in teaching of horsemanship in current and future programs, it speaks both to field officers and riding masters.

Pembroke is unsparing in his criticism of current military horsemanship—"there is a good deal of sense in Xenophon's method of forming horses for war; after him, horsemanship was buried for ages, or rather brutalised, which is still too much the case"—and in his contempt for self-serving riding-masters, incompetent farriers, and dishonest grooms. He spares soldiers and horses, however, because their faults result primarily from ignorance, and not from innate lack of virtue: soldiers and horses are educable and their faults correctible. Abhorring extremes—a "coward" and a "madmen," for example, are alike bad riders—Pembroke urges reason and moderation, patience and gentleness, and simplicity of means as the right tools for educating horses and the soldiers who ride them.

In addition to those basic principles, Pembroke develops recurring themes. He repeatedly notes that scale, urgency, and lack of resources constrain the proper training of cavalrymen and horses, particularly with respect to the teaching of refined techniques. He also repeatedly notes, however, that field maneuvers require precise use of shoulder-in, haunches-in, and rein-back in "opening and closing of files," and that close combat requires agile one-handed riding, since the right hand "carries the sword, which is a sufficient business for it." Finally, as the reader will see, Pembroke has much to say about bitting and shoeing, saddlery and weaponry.

# TYNDALE'S *TREATISE ON MILITARY EQUITATION*

William Tyndale (?-1830), a more elusive biographical subject than Pembroke, was a landowner who inherited an estate at North Cerney upon the death of his father in 1783, and who became Sheriff of Gloucestershire in 1797. As a soldier, Tyndale performed regimental service with appointments as major in the 1st Life Guards in 1794 (exchanged from the 87th Foot), and as brevet lieutenant colonel in 1796. He exchanged to lieutenant-colonel in the 13th Foot in February 1803, and retired from service in August 1803. (His son, Charles William Tyndale, served with notable distinction in the Peninsular War.) Tyndale held an annual pension of 200 pounds as of June 1, 1830, and he died on August 27, 1830.

The first of Tyndale's two tracts, *Instructions for Young Dragoon Officers* (second edition, 1796), is a technical manual. Its first part "instruct[s] the young officer in the part of his duty required in quarters," its second part (four-fifths of the book) "the business of the field"—that is, maneuvers. The first part focuses on discipline, "without which no regiment can possibly be called *good*," the second part on application of that axiom to the field. "When balls are flying round as thick as hail," for example, as Tyndale observes, "System, dressing and all tactick may go to the devil; the quickest way of forming is the best." But "the most regular way [of forming]," he immediately adds, "is the quickest, because it is certain and unconfused." And that way depends on practiced, disciplined soldiers acting in unison "on one principle."

Tyndale's second tract, *A Treatise on Military Equitation* (1797), published in one edition only, starts with the same premise as Pembroke's: "the present system [of horsemanship], which is much the same throughout the cavalry, is contrary to every principle of true horsemanship, both in the instruction of man and horse." Like Pembroke, Tyndale sees little value in training troop horses for the *manège* or troopers in the Art of Equitation when the point is to prepare horse and man for field maneuvers and battle. And, like Pembroke, he argues that young horses and young men must be taught with reason, patience, and simplicity; he stresses the importance of teaching one-handed equitation; and he has much to say about tack and bitting—including the promotion of a military saddle "of my own invention."

Tyndale opens his treatise, however, by invoking Pembroke's *Military Equitation* as "the best work of the kind in our language," but also as a work "too deep and too scientific for a treatise on Military Equitation," as one "beyond the comprehension of those whom it is my wish to teach and reform." Though Tyndale will not presume to match Pembroke's "knowledge of the art," he can perform the service of improving "regimental riding masters" who may know horsemanship through experience but who cannot teach it because they lack knowledge of its underlying principles. Focusing on the *practical application* of those principles of "true horsemanship," Tyndale fixes a narrower objective (and audience) than Pembroke's: it is to teach regimental teachers how to teach.

Tyndale also introduces a common theme in British equestrian writing—the connection between hunting and cavalry training—but does so with a continental bias. The British predilection for hunting, he observes, has caused the military to neglect formal dressage, but hunting offers no preparation for precise maneuvering within "large connected bodies of horsemen." Regimental instructors, in particular, have failed to retrain young recruits in "the method of riding which is necessarily adopted in regiments of cavalry, and which must be adopted by every individual" to prevent confusion—even chaos—in tight formation. Whether Tyndale was right or not, British writers in the next century, ironically, would press the contrary case for hunting as ideal preparation for cavalry.

## APPLICATION

In *The Development of Modern Riding* (1962), the distinguished horseman Vladimir Littauer noted that "Pembroke's is a very small and unimportant book," particularly when compared to the great earlier 18th century masterworks of dressage. Rather than dismiss Pembroke, however, Littauer praised him as "a follower of the French school [of dressage who] was practical enough to suggest its considerable simplification for the army," who successfully repurposed *manège* training for the military work at hand (Littauer, 90-91). What Pembroke did for the French school, in effect, Tyndale did for Pembroke: he further simplified the simplifier.

*What, though, will 21st century amateur equestrians gain from reading two treatises on military equitation written for 18th century professional soldiers?* They will gain historical knowledge of military theory and practice in 18th century England as reflected in the discipline of light cavalry. They will gain historical insight to a moment in the evolution of horsemanship as it was influenced by mounted warfare. And they will find sound principles and lessons to apply to the improvement of their own horsemanship and equitation. In other words, these treatises offer equestrian readers both intellectual and practical value.

Equestrians in virtually any discipline, moreover, can benefit from that value—and eventers most of all. Dressage and military equitation, fox hunting and cavalry service, have been intertwined throughout modern European history. More to the point, three-day eventing, *"le militaire,"* evolved directly from cavalry training of horse and rider. Cavalry officers and chargers, troopers and troop mounts, had to employ the skills later honed by eventers not only to win battles, but also simply to survive them. Not surprisingly, cavalry officers dominated international eventing competition when it developed in the early decades of the 20th century.

Can equestrian readers learn more comprehensive military or equestrian history from current scholarly studies than from 18th century military tracts? Yes. Can readers find more directly applicable lessons in horsemanship and equitation from current training manuals than from those tracts? Yes. Will readers of current works access 18th century minds or hear 18th century voices? No. Will readers witness first hand the deep and enduring human attachment to horses from a historical vantage foreign to ours? Also no. Equestrian readers of Pembroke and Tyndale will engage with two eminently sensible military horsemen, learn from two seasoned trainers of horses and riders, and, if lucky, discover something new and unexpected about equestrian sport.

## NOTES TO THIS EDITION

Bibliographical notes on Pembroke's *A Method of Breaking Horses, and Teaching Soldiers to Ride* and Tyndale's *A Treatise on Military Equitation* following this Introduction are based on the holdings at NSLM. Explanatory notes following each of the two works include proper names, foreign phrases, and a few general terms, but not common terms related to horsemanship, tack and bits, or equine anatomy and care. Where feasible, I have cited 18th century reference works.

*Charles Caramello*

*Professor of English*
*University of Maryland*

*John H. Daniels Fellow*
*National Sporting Library & Museum*

# ACKNOWLEDGMENTS

A John H. Daniels Fellowship at the National Sporting Library and Museum (NSLM) in Middleburg, Virginia, in 2017-2018, enabled this project (together with a sabbatical leave from University of Maryland). I would like to thank NSLM's Board of Directors, Executive Director Melanie Mathewes, and Director of Educational Programs Anne Marie Paquette for their support, and NSLM's librarians, John Connolly and Erica Libhart, for sharing their expertise in NSLM's extraordinary holdings. These holdings include a total of nine copies of first and early editions of Pembroke and Tyndale.

The Editor of Xenophon Press, Richard Williams, supported this project with consistent enthusiasm and good will, and his staff provided the design and production expertise necessary for a handsome and legible facsimile edition with plates photographed from the originals at NSLM. I would like to thank them as well.

# BIBLIOGRAPHICAL NOTES

*A Method of Breaking Horses, and Teaching Soldiers to Ride* was published in four editions during the lifetime of Henry Herbert, 10th Earl of Pembroke. The 1st and 2nd editions (1761 and 1762) appeared under that title; the 3rd and 4th editions (1778 and 1793) under the title, *Military Equitation, Or, A Method of Breaking Horses, and Teaching Soldiers to Ride*. The 1st and 2nd editions were printed in London by J. Hughs, Lincoln's-Inn-Fields; the 3rd edition printed and sold in Sarum [an old name for Salisbury] by E. Easton, and sold also by J. Dodsley, Pall-Mall, and J. Wilkie, St. Paul's Church-Yard, London; and the 4th edition printed for G. and T. Wilkie, No. 57, Pater-Noster-Row, London, and E. and J. Easton, Salisbury.

The 2nd edition, presented as "Revised, and corrected, with Additions," does not differ substantially in contents from the 1st. The 3rd edition, presented as "With Plates. Revised and Corrected, with Additions," includes an added section on bits to Chapter II and one on working horses in hand in Chapter III; Chapter V in the 1st and 2nd editions becomes Chapter VII, and a new Chapter V, *The Trot*, is added; a section on moving pillars is added to Chapter VI; Chapter VII becomes Chapter VIII; Chapter VIII becomes Chapter IX; and an errata list of eight items is added. The 4th edition, presented only as "with plates," does not differ in contents from the 3rd edition.

Two copies of the 1st edition that I have examined each carry one plate depicting an equine foot and horseshoe; two copies of the 2nd edition each carry one plate depicting a bit; a third copy adds a plate depicting a halter. The copies of the 3rd and 4th editions that I have examined each include 17 plates (several of them missing in one copy of the 3rd edition).

Since Pembroke died in 1794, the text of the 4th edition, together with its 17 illustrative plates, can be regarded as authoritative and has been reproduced in the present volume.

*A Treatise on Military Equitation* appeared in one edition during the lifetime of its author, William Tyndale, Lieut. Col. and Major of the First Regiment of Life Guards. Published by subscription in 1793, it was printed for the author and sold by T. Egerton, Military Library, Near Whitehall, London.

The copy that I have examined carries five plates depicting (in order of appearance) a mounted rider, a horse's head, another mounted rider, a diagram of movement in an arena, and a diagram of saddle construction.

The text of this edition, together with its 5 illustrative plates, can be regarded as authoritative and has been reproduced in the present volume.

# MILITARY EQUITATION:

OR,

## A METHOD OF BREAKING HORSES,

AND

## TEACHING SOLDIERS TO RIDE.

*DESIGNED FOR THE USE OF THE ARMY.*

BY
HENRY EARL OF PEMBROKE,
&c. &c. &c.

Scientia, & Patientia.[1]

——— ——— Equitem docuere sub armis
Insultare solo, et gressus glomerare superbos.[2]    VIRG.

Vis consilî expers mole ruit suâ.[3]    HOR.

THE FOURTH EDITION,
WITH PLATES.

LONDON:
PRINTED FOR G. AND T. WILKIE, NO. 57, PATER-NOSTER-ROW;
AND
E. AND J. EASTON, SALISBURY.

MDCCXCIII.

# TO THE KING.

SIR,

WHEN the first regiment of light dragoons was raised under the command of my friend General GEORGE AUGUSTUS ELIOTT,[4] we had frequent occasion to lament together the wretched system of HORSEMANSHIP, that at present prevails in the ARMY: A system disgraceful in itself, and productive in its consequences of the most fatal evils: For troops in their own nature most excellent and brave have been frequently rendered inferior to less powerful ones, both in men and horses, for want of proper instructions and intelligence in this Art. These serious considerations (for indeed they are very much so) induced me to write down and make public the following Lessons, calculated for the use of the Cavalry: They are such as I have always practised

## DEDICATION.

tifed myfelf; and taught both in the above-mentioned regiment and elfewhere, with conftant fuccefs. Incited by thefe reafons, I thus prefume to lay at your Majefty's feet this little work, the outlines only of a more extenfive, general one, which I intend to make public hereafter, fhould I find time to finifh it: And I am the more encouraged to it from the honour You have often done me of talking to me upon HORSEMANSHIP, as alfo from this confidence, that if what I here recommend, be deemed in any wife likely to be ufeful, (as I flatter myfelf it may, if candidly examined, and judicioufly practifed) it will not fail of receiving Your MAJESTY's Royal Approbation and Support. I am,

SIRE

YOUR MAJESTY's

MOST DUTIFUL SUBJECT,

AND DEVOTED SERVANT,

PEMBROKE.

PEMBROKE-HOUSE,
FEB. 15, 1761.

# CONTENTS OF THE FOLLOWING TREATISE.

### CHAP. I.

*The method of preparing horses to be mounted, with the circumstances relative to it.* - - - - page 1

### CHAP. II.

*The method of placing the men and rendering them firm on horseback, with some occasional instructions for them and the horses; and of bits.* - - - page 6

### CHAP. III.

*The method of suppling horses with men upon them, by the* Epaule *en dedans, &c. with and without a longe, on circles and on strait lines; and of working horses in hand.* - - - - - - - - - - page 31

CHAP.

# CONTENTS.

### CHAP. IV.

*Of the head to the wall, and of the croupe to the wall.* — page 53.

### CHAP. V.

*The Trot.* — page 61.

### CHAP. VI.

*The method of reining back---and of moving forwards immediately after---of piaffing---of pillars, &c.---of moving pillars, &c.* — page 71.

### CHAP. VII.

*The method of teaching horses to stand fire, noises, alarms, sights, &c.---of preventing their lying down in the water---to stand quiet to be shot off from---to go over rough and bad ground---to leap hedges, gates, ditches, &c. standing and flying---to disregard dead horses---to swim, &c.* — page 80.

CHAP.

# CONTENTS.

## CHAP. VIII.

*The method of curing restivenesses, vices, defences, starting, and stumbling, &c.* - - - - - page 88

## CHAP. IX.

*Several remarks and hints on shoeing, feeding, management of horses, &c. &c.* - - - - - page 97

---

### ERRATA.

Page 42. l. 14. dele the *comma* after *properly*.
    57. l. 15. for *appui's* read *appuis*.
    61. l. 4. place a *comma* after *determine*.
    69. l. 2. place a *colon* after *themselves*.
    90. l. 11. place a *full stop* after *to*.
  101. l. 15. for *detestible* read *detestable*.
  103. l. 17. for *bevill'd* read *bevelled*.
  120. l. 17. dele *on* after *almost*.

# A METHOD OF BREAKING HORSES,

### AND TEACHING SOLDIERS TO RIDE, &c.

## CHAP. I.

*The method of preparing horses to be mounted, with the circumstances relative to it.*

THOUGH all horses for the service are generally bought at an age, when they have already been backed, I would have them begun and prepared for the rider with the same care, gentleness and caution, as if they had never been handled or backed, in order to prevent accidents, which might else arise from skittishness or other causes: and as it is proper, that they should be taught the figure of the ground they are to go upon, when they are at first mounted, they should be previously trotted in a *longe* on large circles, without any one upon them, and without a saddle, or any thing else, at first, which might hurt, constrain, tickle, or make them any ways uneasy.

The manner of doing this is as follows: Put an easy *cavesson* upon the horse's nose, and make him go forwards round you, standing quiet and holding the *longe*; and let another man, if you find it necessary, follow him with a whip: All this must be done very gently, and but a little at a time; for more horses are spoilt by over-much work, than by any other treatment whatever; and that by very contrary effects, for sometimes it drives them into vice, madness and despair, and often it stupifies them and totally dispirits them. An excellent way of *longing* horses, who are apt to carry their heads low, (which many do) is to *longe* them with a cord buckled to the top of the head-stall, and passing from thence through the eye of the snaffle into the hand of the person who holds the *longe*.

The first obedience required in a horse, is going forwards: 'Till he performs this duty freely, never even think of making him rein back, which would inevitably render him restive: As soon as he goes forwards readily, stop and caress him. You must remember in this, and likewise in every other exercise, to use him to go equally well, to the right and left; and when he obeys, caress him and dismiss him immediately. A horse, though ever so perfect to one hand only, is but a half dressed horse. If a horse, that is very young, takes fright and stands still, lead on another horse before him, which probably will induce him instantly to follow. Put a snaffle in his mouth; which snaffle should be full, and thick in the mouth-piece, and not too short: and when he goes freely, saddle him, girting him at first very loose. Let the cord, which you hold, be long and loose; but not so much so, as to endanger the horse's entangling his legs in it. It must be observed, that small circles, in the beginning, would constrain the horse too much, and put him upon defending himself. No bend must be required at first: never suffer him to gallop false; but whenever he attempts it, stop him without delay, and then set him off afresh. If he gallops of his own accord, and true, permit him to continue it; but if he does it not voluntarily, do not demand it of him at first. Should he fly and jump, shake the cord gently upon his nose without jerking it, and he will fall into his trot again. If he stands still, plunges or rears, let the man, who holds the whip, make a noise with it; but never touch him, 'till it be absolutely necessary to make him go on. When you change hands, stop and caress him, and entice him by fair means to come up to you: for by presenting yourself, as some do, on a sudden before horses, and frightening them to the other side, you run a great risk of giving them a shyness. If he keeps his head too low, heighten your hand, and shake the *cavesson* to make him raise it: And in whatever the horse does, whether he walks, trots, or gallops, let it be a constant rule, that the motion be determined and really such as is intended, without the least shuffling, pacing, or any other irregular gait. A false gait should never be suffered. The trot is the pace, which enables all quadrupeds to balance and support themselves with firmness and ease. When he goes lightly, and freely,

tie

tie his head a little inwards by degrees: more, and more so, as he grows more supple, both in trotting, and galloping, in the *longe*, without any one upon him. Great care must be taken, that he always goes true, and that his head is not kept tied for any time together; for if it was, he would infallibly get a trick of leaning on the rein, and throw himself heavily upon his shoulders, when he grew tired. Every regiment should have some covered place for their riding during the winter, or nothing hardly can be done in the bad season. In good weather, it is full as well, and more pleasant, to work out of doors: and indeed doing so frequently prevents local routines, which horses are sometimes particularly apt to take in shut schools, if great care is not taken. On the other hand, they are more often *distraied*, and apt to lose their attention by various objects, in fields, than they are in a riding-house. It is therefore difficult to decide, either for the one, or the other. There is more liberty in the one, than in the other, and horses out of doors grow used to objects they would otherwise fear. In shut schools, work may be more exactly done, perhaps, and the ground there is best. Both are good at proper seasons, and either will do very well, if the Riding-Master is good.

CHAP.

## CHAP. II.

*The method of placing the men, and rendering them firm on horseback; with some occasional instructions for them and the horses; and of bits.*

'TIS necessary that the greatest attention, and the same gentleness, that is used in teaching the horses, be observed likewise in teaching the men, especially at the beginning. Every method and art must be practised to create and preserve, both in man and horse, all possible feeling and sensibility, contrary to the usage of most riding-masters, who seem industriously to labour at abolishing these principles both in the one and the other. As so many essential points depend upon the manner, in which a man is at first placed on horseback, it ought to be considered, and attended to with the strictest care and exactness.

The absurdity of putting a man, who perhaps has never before been upon a horse, (or if he has, 'tis probably so much the worse) on a rough trotting one, on which he is obliged (supposing the horse is insensible enough to suffer it; and if he be not, the man runs a great rifk of breaking his neck) to ftick with all the force of his arms and legs, is too obvious to need mentioning. This rough work, all at once, is plainly as detrimental at firft, as it is excellent afterwards in proper time. No man can be either well, or firmly feated on horfeback, unlefs he be mafter of the ballance of his body, quite unconftrained, with a full poffeffion of himfelf, and at his eafe, on all occafions whatever; none of which requifites can he enjoy, if his attention be otherwife engaged; as it muft wholly be in a raw, unfuppled, and unprepared lad, who is put at once upon a rough horfe: In fuch a diftrefsful ftate he is forced to keep himfelf on at any rate, by holding to the bridle, (at the expence of the fenfibility both of his own hand, and the horfe's mouth) and by clinging with his legs, in danger of his life, and to the certain depravation of a right feeling in the horfe;---a thing abfolutely neceffary to be kept delicate, for the forming properly both of man and horfe; not to mention the horrid appearance of fuch a figure, rendered totally incapable of ufe and action.

The firft time a man is put on horfeback, it ought to be

be upon a very gentle one. He never should be made to trot, 'till he is quite easy in the walk, and then on very easy horses at first. Afterwards, as he grows firmer, put him on rougher horses, and augment by degrees the velocity of the trot. He should not gallop, 'till he can trot well; because, though the motion of the gallop is the easiest, a horse may be more easily unsettled in galloping than in trotting. The same must be observed in regard to horses: they should never be made to trot, 'till they are obedient, and their mouths are well formed on a walk; nor be made to gallop, 'till the same be effected on a trot. When he is arrived at such a degree of firmness in his seat, the more he trots, (which no man whatever should ever leave off) and the more he rides rough horses, the better. This is not only the best method, (I may say, the only right one) but also the easiest and the shortest: by it, a man is soon made sufficiently an horseman for a soldier; but by the other detestable methods, that are commonly used, a man, instead of improving, contracts all sorts of bad habits, and rides worse and worse every day; the horse too becomes daily more and more unfit for use. In proceeding according to the manner I have propofed, a man is rendered firm and eafy upon the horfe, and, as it were, of a piece with him; both his own and the horfe's fenfibility is preferved, and each in a fituation fit to receive and practife all leffons effectually: for if the man and horfe do not both work without difficulty and conftraint, the more they are exercifed, the worfe they become; every thing they do, is void of all grace, and of all ufe. When the man has acquired a perfect firmnefs on a faddle, he fhould by degrees be made equally firm on a rug, or on a horfe's bare back; fo much fo, as to be as firm, to work as well, and be quite as much at his eafe, as on any demi-pique faddle. Very little patience and attention will bring this about.

Among the various methods, that are ufed, of placing people on horfeback, few are directed by reafon. Some infift, that fcarce any preffure at all fhould be upon the backfide; others would have the feat be almoft upon the back-bone: out of thefe two contrary, and equally ridiculous methods, an excellent one may be found, by taking the medium. Before you let the man mount, teach him to know, and always to examine, if the curb be well placed,

placed, (I mean, when the horse has a bit in his mouth, which at first he should not, but only a snaffle, 'till the rider is firm in his seat, and the horse also somewhat taught) and likewise if the nose-band be properly tight; the throat-band loosish, and the mouth-piece neither too high, nor too low in the horse's mouth, but rightly put, so as not to wrinkle the skin, nor to hang lax; the girts drawn moderately, but not too tight; the crupper, and the breast-plate, properly adjusted, and whether the reins are of equal length. They should be frequently taken off and made so, when they are found not to be so. A very good and careful hand may venture on a bit at first, and succeed with it full as well, as by beginning with a snaffle alone: but such a proceeding will require more care, more delicacy, and more time, than can be expected in a corps, whose numbers are so considerable, and where there are so few, if any good riders: A raw man is much easier taught to do well, than one, who has learnt ever so long, on bad principles; for it is much more difficult to undo, than to do; and the same in respect to the horse. On colts, it is better in all schools whatsoever, to avoid any pressure on the bars just at first, which a curb, though ever so delicately

cately used, must in some degree occasion. Whoever begins a horse with a bridle, must be, in every respect, a very good, delicate rider, and be very careful that the horse does not get and keep his head low, whereby all action in the shoulders is spoiled. I have seen some schools, in France particularly, where a bit was immediately put into a horse's mouth at first; but I have constantly observed in those schools, that their horses carried their heads low, that the motion of their shoulders was not free, but confined. Here and there one horse or so, indeed, there might be, whose fore-hand nature had placed so high, that nothing could bring it down low. Great care must be taken to make the men use their snaffles delicately; otherwise, as a snaffle has not the power, which a bridle has upon a horse's mouth, they will use themselves to take such liberties with it, as will quite spoil their hands, and teach the horses to pull, be dead in hand, and quite upon their shoulders, entirely deprived of good action. Whenever any bridles are used, (and they always should be at a proper time, when the horses' heads are high, and they are well determined, light in hand, and free in their motions) they must be all the same; for though different mouths require different sorts of bits, it is absolutely necessary that some general uniform sort should be used throughout a whole regiment. They should differ only in breadth, according to the breadth of each horse's mouth. There needs no great variety of sizes for bitting a whole regiment. The best I could ever pitch on, after repeated trials, is one made after the following drawing. (*Plate* 1.) The weight of the bit, without the curb, is about fourteen ounces three quarters, the curb alone weighs about four ounces and a quarter, and the little chain to prevent horses taking the branches in their mouth, (which is a trick very many horses get) three quarters of an ounce. The whole together weighs one pound, three ounces, and three quarters. The rings to the branches should be fixed, and the reins buckled to them, to prevent the latter from twisting. The mouth-piece is of a proper shape, height, and substance, and is fixed. All such as are not so, and move in the joint, have a bad, uncertain effect. Thin curbs are bad, and apt, if at all roughly used, (a thing very difficult to prevent at all times in some people's hands) to cut, and damage the horse's mouth very much. They should be flat, broad,

and

and easy, that they may not hurt the horse's *barbe*, but they must not be thick, or heavy. This bridle is calculated for light troops. Heavier corps, who have larger horses, and of another kind, may have the branches a quarter of an inch longer, and the whole bridle somewhat, but very little more substantial. Bridles should never be used with raw recruits, or with raw horses, at first: a plain mouthed, smooth snaffle, does much better; the twisted, sharp, cutting ones, are barbarous, callous making instruments at best; the single ones, as well as the double rein ones, are often very useful, and agreeable even with dressed horses upon all airs whatsoever, if they are apt to get their heads low. When these necessary precautions have been all taken, let the man approach the horse gently near the shoulder; then taking the reins and an handful of the mane in his left hand, let him put his left foot softly into the left stirrup, (but not too far in) by pulling it towards him, lest he touch the horse with his toe, which might frighten him; then raising himself up, let him rest a moment on it with his body upright, but not stiff: and after that, passing his right leg clear over the saddle, without rubbing against any thing, let him seat

himself gently down. The same precautions must also be taken in dismounting. He must be cautious not to take the reins too short, for fear of making the horse rear, run, or fall back, or throw up his head; but let him hold them of an equal length, neither tight nor slack, and with the little finger betwixt them. 'Tis fit that horses should be accustomed to stand still to be mounted, and not stir 'till the rider pleases. The man, who holds the horse to be mounted, must not do it by the bridle, but only by the cheeks of the head-stall, and gently, otherwise the same inconvenience might arise, as from the rider's holding the reins too short himself in mounting. All soldiers should be instructed to mount and dismount equally well on both sides, which may be of very great use in times of hurry and confusion. Place the man in his saddle, with his body rather back, and his head held up with ease, without stiffness; seated neither forwards, nor very backwards, with the breast pushed out a little, and the lower part of the body likewise a little forwards; the thighs and legs turned in without constraint, and the feet in a strait line, neither turned in nor out: By this position, the natural weight of the thighs has a proper and sufficient pressure of itself,

ſelf, and the legs are in readineſs to act, when called upon: they muſt hang down eaſy and naturally, and be ſo placed, as not to be wriggling about, touching and tickling the horſe's ſides, but always near them in caſe they ſhould be wanted, as well as the heels.

The body muſt be carefully kept eaſy and firm, and without any rocking, when in motion; which is a bad habit very eaſily contracted, eſpecially in galloping. The left elbow muſt be gently leant againſt the body, a little forwards; unleſs it be ſo reſted, the hand cannot be ſteady, but will be always checking, and conſequently have pernicious effects on the horſe's mouth: and the hand ought to be of equal height with the elbow; if it were lower, it would conſtrain and confine the motion of the horſe's ſhoulders, which muſt be free. I ſpeak here of the poſition of the hand in general; for as the mouths of horſes are different, the place of the hand alſo muſt occaſionally differ: a leaning, low, heavy fore-hand, requires a high hand; and a horſe that pokes out his noſe, a low one. The right hand arm muſt be placed in ſymmetry with the left; only let the right hand be a little forwarder

warder or backwarder, higher or lower, as occasions may require: in order that both hands may be free, both arms must be a little bent at the elbow, to prevent stiffness.

A soldier's right hand should be kept unemployed in riding; it carries the sword, which is a sufficient business for it: In learning therefore to ride, the men should have a whip or switch in it, and hold it upwards, that they may thereby know how to carry their swords properly, keeping it downwards only, when they mount or dismount, that the horse may not be frightened at the sight of it.

The hand must be kept clear of the body, about two inches and a half forwards from it, with the nails turned opposite to the waistcoat buttons, and the wrist a little rounded with ease; a position not less graceful than ready for slackening, tightening, and moving the reins from one side to the other, as may be found necessary.

A firm and well balanced position of the body, on horseback, is (as has already been said) of the utmost consequence; as it affects the horse in every motion, and is the best of helps: whereas on the contrary, the want of it is the greatest detriment to him, and an impediment in all his actions. Many people make a great difference about saddles, as a serious object of firmness; but nobody can be truly said to have a seat, who is not equally firm on flat, or demi-piqued saddles,[5] on the true principles of equilibre, and ease. When the men are well placed, the more rough trotting they have, without stirrups, the better; but with a strict care always, that their position be preserved very exactly. As for those unfeeling fellows, who continue sticking by their hands, in spite of all the teacher's attention to prevent it, nothing remains to be done, but to make them drop the reins quite on a safe-going horse, and to keep their hands in the same position, as if they held them. In all cases without exception, but more especially in this, great care must be taken to hinder their clinging with their legs: in short, no sticking by hands or legs is ever to be allowed of at any time. If the motion of the horse be too rough, slacken it, 'till the rider grows by degrees more firm: and when he is quite firm and easy on his horse in every kind of motion, stirrups may be given him; but he must never leave off trotting often, and working often without any.

The stirrups must be neither short nor long; but of such a length that when the rider, being well placed, puts his feet into them, (about one-third of the length of the foot from the point of it) the points may be between two and three inches higher than the heels: longer stirrups are bad, and would make it very difficult for the rider to get his leg over the baggage, forage, cloak, &c. which are fastened on behind upon the saddle: and shorter would be bad in every respect, and answer no end at all. The length I mentioned above, is just the right one, and is to be taken in the following method: make the rider place himself upon the saddle, even, upright and well, with his legs hanging down, and the stirrups likewise: and when he is in this position, raise the rider's toe to an equal height with his heel, and take up the stirrup, 'till the bottom of it comes just under the ankle-bone. The stirrups must be exactly of an equal length. The rider must not bear upon his stirrups, but only let the natural weight of his legs rest on them: for if he bore upon them, he would be raised above, and out of his saddle; which should never be, except in charging sword in hand, with the body inclined forwards at the very instant of attacking.

ing. Spurs may be given, as soon as the rider is grown familiar with stirrups, or even long before, if his legs are well placed.

Delicacy in the use of the hands, as well as in the use of the legs, may be given by the teacher to a certain degree; but 'tis nature alone that can bestow that great sensibility, without which neither one nor the other can be formed to any great perfection. A hand should be firm, but delicate: a horse's mouth should never be surprised by any sudden transition of it, either from slack to tight, or from tight to slack. Every thing in horsemanship must be effected by degrees, and with delicacy, but at the same time with spirit and resolution. That hand, which by giving and taking properly, gains its point with the least force, is the best; and the horse's mouth, under this same hand's directions, will also consequently be the best, supposing equal advantages in both from nature. This principle of gentleness should be observed upon all occasions in every branch of horsemanship. Hard, bad mouths, may appear soft and good to an insensible hand; so that it is impossible to form any judgment of a horse's mouth by what any body tells you of it, unless you know the degree of sensibility, and science that person is possessed of in horsemanship, or ride the horse yourself. Sometimes the right hand may be necessary, for a moment, upon some troublesome horses, to assist the left; but the seldomer this is done, the better; especially in a soldier, who has a sword to carry, and to make use of,

The snaffle must on all occasions be uppermost, that is to say, the reins of it must be above those of the bridle, whether the snaffle or the bit be used separately, or whether they be both used together. When the rider knows enough, and the horse is sufficiently prepared and settled to begin any work towards suppling, one rein must be shortened according to the side worked to, (as is explained in its proper place) but it must never be so much shortened, as to make the whole strength rest on that rein alone; for, not to mention that the work would be false and bad, one side of the horse's mouth would by that means be always deadened; whereas on the contrary, it should always be kept fresh by its own play, and by the help of the opposite rein's acting delicately in a smaller degree

degree of tenſion; the joint effects of which produce in a horſe's mouth the proper, gentle, and eaſy degree of *appui*[6] or bearing; to preſerve which, when obtained, the horſe muſt not be over-worked; if he is, he will, beſides other bad conſequences, throw himſelf on his ſhoulders into the rider's hand, like a tired poſt-horſe on the road. Colts indeed, as well as men, at firſt muſt be taught the effect of the reins taken ſeparately, for fear of confounding them in the beginning with mixed effects of them at once. Avoid working in deep, bad ground; beſides its ſpoiling a horſe's paces, it obliges him to throw himſelf on his ſhoulders upon the rider's hand, and teaches him to toſs his head about diſagreeably.

A coward and a madman make alike bad riders, and are both alike diſcovered and confounded by the ſuperior ſenſe of the creature they are mounted upon, who is equally ſpoilt by both, though in very different ways. The coward, by ſuffering the animal to have his own way, not only confirms him in his bad habits, but creates new ones in him: and the madman, by falſe and violent motions and corrections, ruins the horſe, and drives him,

through

through despair, into every bad and vicious trick that rage can suggest.

All horses heads must be kept very high, 'till they are quite determined, and free in the motions of their shoulders.

It is very requisite in horsemanship, that the hand and legs should act in correspondence with each other in every thing; the latter always subservient and assistant to the former. Upon circles, in walking, trotting, or galloping, (I mean only where nothing more is intended) the outward leg is the only one to be used, and that only for a moment at a time, in order to make the horse go true, if he be false; and as soon as that is done, it must be taken away again immediately. If the horse is lazy, or any ways retains himself, both legs must be used, and pressed to his sides at the same time together; if after having tried softer methods, such as a gentle pressure of the thighs, and putting the legs back, they should fail, but not before. The less the legs are used in general, the better. Very delicate riders, in regular well attended good schools, never want their help; and horses so dressed, fed, are by far superior to all others: they obey the smallest touch on the rein, or the least weight of the body thrown one way, or the other, imperceptibly, as may be necessary: the horse and man seem one, and the same, and such is the practice and teaching of great masters; but that perfection in the feeling of either man, or horse, is not to be expected in the hurry which can not be avoided in a regimental school, where the numbers are so great. By the term outward, is understood the side which is more remote from the center; and by inward, is meant the side next to the center. In reining back, the rider should be careful not to use his legs, unless the horse backs on his shoulders; in which case, they must be both applied gently at the same time, and correspond with the hand. If the horse refuse to back at all, the rider's legs must be gently approached, 'till the horse lifts up a leg, as if to go forwards; at which time, when that leg is in the air, the rein of the same side with that leg, which is lifted up, will easily bring that same leg backwards, and accordingly oblige the horse to back: but if the horse offers to rear, the legs must be instantly removed away. The inward rein must be the tighter on circles, so that

the

the horse may bend and look inwards; and the outward one crossed over a little towards it; and both held in the left hand, that soldiers may not have their right employed, which, as has before been observed, must be left free for other more necessary uses.

Let the man and horse begin all lessons whatsoever on very slow motions, that they may have time to understand, and reflect on what is taught them; but though the motions are slow, they must not be dull, but determined, and without hesitation. In proportion as the effects of the reins are better comprehended, and the manner of working becomes more familiar, the quickness of motion must be increased. Every rider must learn to feel, without the help of the eye, when a horse goes false, even in the most speedy, and most violent motions, and remedy the fault accordingly: this is an intelligence, which nothing but practice, application, and attention, can give, in the beginning on slow motions. A horse may not only gallop false, but also trot and walk false. If a horse gallops false, that is to say, if going to the right, he leads with the left leg; or if going to the left, he leads with the right;

right; or in case he is disunited, by which is meant, if he leads with the opposite leg behind to that which he leads with before, stop him immediately, and put him off again properly: the method of effecting this, is by approaching your outward leg, gently, and putting your hand outwards, still keeping the inward rein the shorter, and the horse's head inwards, if possible; but if he should still resist, then bend and pull his head outwards also. Replace it again, bent properly inwards, the moment he goes off true. The help of the leg in this, and indeed all other cases, must not be made use of at all, 'till that of the hand alone has proved ineffectual. A horse is said to be disunited to the right, when going to the right, and consequently leading with the right leg before, he leads with the left behind; and is said to be disunited to the left, when going to the left, and consequently leading with the left leg before, he leads with the right behind. A horse may at the same time be both false and disunited; in correcting both which faults, the same method must be used. He is both false and disunited to the right, when in going to the right he leads with the left leg before, and the right behind; notwithstanding that hinder leg be with propriety more forward under his belly, than the left, because the horse is working to the right: and he is false and disunited to the left, when in going to the left, he leads with the right leg before, and the left behind; notwithstanding, as above, that hinder leg be with propriety more forward under his belly than the right, because the horse is working to the left.

Care must be taken, that horses, in stopping on the gallop, stop true, behind particularly, which they are very apt not to do; especially in the longe, and bent, without any one on them.

In teaching men a right seat on horseback, the greatest attention must be given to prevent stiffness, and sticking by force in any manner upon any occasion: stiffness disgraces every work; and sticking serves only to throw a man (when displaced) a great distance from his horse, by the spring he must go off with: whereas by a proper equilibrating position of the body, and by the natural weight only of the thighs, he cannot but be firm, and secure in his seat.

As the men become more firm, and the horses more supple, 'tis proper to make the circles less, but not too much so, for fear of throwing the horses forwards upon their shoulders.

No bits should be used, 'till the riders are firm, and the horses bend well to right and left; and then too always with the greatest care and gentleness. The silly custom of using strong and heavy bits, is in all good schools with reason laid aside, as it should be likewise in military riding: they pull down the horse's head, keep it low, thereby obstruct the action of the fore parts, and harden as much the hand of the rider, as the mouth of the horse; both which becoming every day more and more insensible together, nothing can be expected but a most unfeeling callousness both in one and the other. Some horses, when first the bit is put into their mouths, if great care be not taken, will put their heads very low; which low position of the head, provided the top of the head, and the nose, be nearly perpendicular, some ignorant people call a good one; without considering, that the higher the top of the head is, provided that it is nearly perpendicular with the nose, the better the position is on every account. If the top of the head is low, the position is a bad one, notwithstanding the head and nose being nearly perpendicular, because it obstructs the action of the fore parts. With such horses, raise your right hand with the *bridoon* in it, and play at the same time with the bit in the left hand, giving and taking. A strong bit, indeed, will flatter an ignorant hand, just at first; but it will never any other, nor even an ignorant one for any time together; for the horse's mouth will soon grow callous to it, and unfeeling, and the hand the same. Most horses, whose heads are heavy, are apt to stumble.

On circles, the rider must lean his body inwards; unless great attention be given to make him do it, he will be perpetually losing his seat outwards, every rapid or irregular motion the horse may make. 'Tis scarce possible for him to be displaced, if he leans his body properly inwards.

Instructions, both to man and horse, in riding, are of the greatest importance and consequence; as the success of actions in a great measure depends upon them. Squadrons are frequently broken and defeated through the ignorance of the riders, or horses, but most commonly of both together. Many and various are the disasters, that arise from the horses not being properly prepared and suppled, and from the men not being taught firm seats, independent of their hands, and the mouths of their horses. Were the men rightly instructed how to keep the mouths of their horses fresh and obedient, and thereby maintain a cadenced pace, (be it ever so fast, or ever so slow) ranks would of course be always dressed, and unshaken, and consequently always powerful. The stoutest, and by nature, the best of cavalry, is often broken, and thereby rendered inferior far to much weaker and less respectable bodies than themselves, for want of being properly informed in the above-mentioned, and such-like particulars. This is a matter worthy of a serious inspection, consideration, and amendment, the neglect of which has upon many occasions been very fatal. 'Tis to be hoped, that some person of sufficient authority and knowledge will contrive to introduce many alterations, that appear very necessary in the cavalry. To what purpose is cavalry loaded with such monstrous heavy boots and firelock? a lighter,

yet

yet full as ſtrong, and much more ſerviceable boot might be eaſily contrived. A light carabine⁷ would ſuit them far better. A hat ſeems to me a ſilly and uſeleſs piece of dreſs in a ſoldier: it is continually falling off, eſpecially in action; nor can it ever ſerve as a protection againſt blows, &c. or bad weather, which are circumſtances of great conſequence: whereas a cap has no inconveniences at all attending it, may be made very ornamental and of a martial appearance, and in ſuch a manner, as to be a good fence againſt blows, rain, ſnow, and ſtormy winds, and alſo convenient to ſleep in.

## CHAP. III.

*The method of suppling horses, with men upon them, by the* EPAULE *en dedans, &c. with and without a* longe, *on circles and on strait lines; and of working horses in hand.*

WHEN a horse is well prepared and settled in all his motions, ('till when nothing more must be attempted) and the rider firm, (which is also as absolutely necessary) it will be proper then to proceed on towards a farther suppling and teaching of both. In regiments, especially those that are young, there are but very few, if any, tolerable horsemen; which makes the greatest exactness and gentleness absolutely necessary in the instructing of both: and more particularly so in this case, as horse and man are both ignorant, and must be both alike taught together; which is a difficulty, that does not exist in schools; for there a young rider is put upon a made, or at least a quiet horse; nor do any, but able riders, ever mount a raw one.

In setting out upon this new work, before which the horse should be taught to go well into the corners, both with his fore and hinder parts, on a walk, (without being bent, for that cannot be yet expected, though it will be soon) and be very light in hand; when he does it, begin by bringing the horse's head a little more inwards than before, pulling the inward rein gently to you by degrees. When this is done, try to gain a little on the shoulders, by keeping the inward rein the shorter, as before, and the outward one crossed over towards the inward one. The intention of these operations is this: the inward rein serves to bring in the head, and procures the bend; whilst the outward one, that is a little crossed, tends to make that bend perpendicular, and as it should be; that is to say, to reduce the nose and the forehead to be in a perpendicular line with each other: it also serves, if put forwards, as well as also crossed, to put the horse forwards, if found necessary; which is often requisite, many horses being apt in this, and other works, rather to lose their ground backwards, than otherwise, when they should rather advance: if the nose were drawn in towards the breast beyond the perpendicular, it would confine the motion

tion of the shoulders, and have other bad effects. All other bends, besides what I have above specified, are false. The outward rein, being crossed, not in a forward sense, but rather a little backwards, serves also, when necessary, to prevent the outward shoulder from getting too forwards, which facilitates the inward leg's crossing it; which is the motion that so admirably supples the shoulders. Care must be taken, that the inward leg pass over the outward one, without touching it; this inward leg's crossing over must be helped by the inward rein, which you must cross towards and over the outward rein, every time the outward leg comes to the ground, in order to lift and help the inward leg over it: at any other time, but just when the outward leg is come to the ground, it would be wrong to cross the inward rein, or to attempt to lift up the inward leg by it: nay, it would be demanding an absolute impossibility, and lugging about the reins and horse to no purpose; because a very great part of the horse's weight resting upon the inward leg would render such an attempt, not only fruitless, but also prejudicial to the sensibility of the mouth, and probably oblige him to defend himself, without being productive of any suppling motion whatsoever.

When the horse is thus far familiarly accustomed to what you have required of him, (but by no means before he is entirely so) then proceed to effect by degrees the same crossing in his hinder legs. By bringing in the fore legs more, you will of course engage the hinder ones in the same work: if they resist, the rider must bring both reins more inwards; and, if necessary, put back also, and approach his inward leg to the horse: and if the horse throws out his croup too far, the rider must bring both reins outwards, and if absolutely necessary, (but not otherwise) he must also delicately make use of his outward leg for a moment; in order to replace the horse properly; observing, that the croup should always be considerably behind the shoulders, which in all actions must go first; and the moment that the horse obeys, the rider must put his hand and leg again into their usual position. In this lesson, as indeed in almost all others, the corners must not be neglected: the horse should go well, and thoroughly into them. Bring his fore parts into them, by

crossing

crossing over the inward rein towards the outward one, (but without taking off from the proper bend of the head, neck, and shoulders) and bring them out of the corner again by crossing over the outward rein towards the inward one. These uses of the reins have also their proper effects upon the hinder parts.

Nothing is more ungraceful in itself, more detrimental to a man's seat, or more destructive of the sensibility of a horse's sides, than a continual wriggling unsettledness in a horseman's legs, which prevents the horse from ever going a moment together true, steady, or determined. 'Tis impossible, upon the whole, for a man to be too firm, settled, and gentle. A soft motion may be always inforced, if necessary, with ease; but an harsh one is irrecoverable, and its bad consequences very often almost irreparable. Men are very apt to get this trick of wriggling their legs, even in going strait forward, and more so with one leg particularly put back in changing of hands; which should be done by the reins only, in a graceful, still manner, and without letting the horse either throw himself over too fast, or go lazily over to the other hand:

the rider's hand alone is almoſt always ſufficient; and, if it ſhould not, many things ſhould be tried, before ſo ugly, and bad a reſource, as the above-mentioned is thought of; 1ſt, that of ſqueezing the thighs; 2d, approaching gently the calves of the legs, and 3d, uſing the ſpur; but without diſtorting the leg, or foot, which a good maſter will not permit to be done.

A horſe ſhould never be turned, without firſt moving a ſtep forwards; an imperceptible motion only of the hand, from one ſide to the other, is ſufficient to turn him. It muſt alſo be a conſtant rule, never to ſuffer a horſe to be ſtopped, mounted, or diſmounted, but when he is well placed.

At firſt, the figures worked upon muſt be great, and afterwards made leſs by degrees, according to the improvement which the man and horſe make; and the cadenced pace alſo, which they work in, muſt be accordingly augmented. The changes from one ſide to the other, muſt be in a bold, determined trot, and at firſt quite ſtraight forwards, without demanding any ſide motion on two *piſtes*, which it is very neceſſary to require afterwards,

when

when the horse is sufficiently suppled. By two *pistes* is meant, when the fore parts and hinder parts do not follow, but describe two different lines.

In the beginning, a *longe* is useful on circles, and also on straight lines, to help both the rider and the horse; but afterwards, when they are grown more intelligent, they should go alone. No one, not even the best riders, should ever quite leave off trotting every now and then, in the *longe*, both with, and without stirrups. At the end of the lesson rein back, and then put the horse, by a little at a time, forwards, by approaching both legs gently, and with an equal degree of pressure, to his sides, (if necessary) and playing with the bridle: if he rears, push him out immediately into a full trot. Shaking the *cavesson* on the horse's nose, and also putting one's self before him, and rather near to him, will generally make him back, though he otherwise refuse to do it: and moreover, a slight use and approaching of the rider's legs, will sometimes be necessary in backing, in order to prevent the horse from doing it too much upon his shoulders; but the pressure of the legs ought to be very small, and taken

quite

quite away the moment that he puts himself enough upon his haunches. The horse must learn by degrees to back upon a straight line, but to make him do so, the rider must not be permitted to have recourse immediately to his leg, and so distort himself by it, (which is generally practised with the common sort of riding-masters) but first try, if crossing over his hand and reins, to which ever side may be necessary, will not be alone sufficient; which most frequently it will; if not, then employ the leg, which should never be used 'till the last extremity.

After a horse is well prepared, and settled, and goes freely on in all his several paces, he ought to be in all his works kept, to a proper degree, upon his haunches, with his hinder legs well placed under him; whereby he will be always pleasant to himself, and his rider, will be light in hand, and ready to execute whatever may be demanded of him in reason, with facility, vigour, quickness, and delicacy.

The common method, that is used, of forcing a horse sideways, is a most glaring absurdity, and very hurtful to the animal in its consequences; for, instead of suppling

pling him, it obliges him to stiffen and defend himself, and often makes a creature, that is naturally benevolent, a restive, frightened, and vicious man-hater for ever. In general 'tis a maxim, as constantly to be remembered, as it is true, that it is more difficult to correct faults and bad habits, than to foresee and prevent them. Horses under riders, who use their legs, are, when going to work on two pistes, perpetually setting off with the croup foremost, than which nothing hardly can be worse. It is owing to the leg of the rider being applied to the side of the horse, before the hand has determined the fore parts of the animal, on the line, upon which he is to go.

For horses, who have very long and high fore-hands, and who poke out their noses, a running snaffle is of excellent use; but for such, as bore and keep their heads low, a common one is preferable; though any horse's head indeed may be kept up also with a running one, by the rider's keeping his hands very high and forwards; but that occasions a bad and aukward position in the man. They are, as plainly appears from their construction, bad for tripping and stumbling horses. Whenever either is

used

used alone, without a bridle, upon horses that carry their heads low, and that bore, it must be gently sawed about from one side to the other.

Every body knows the construction of a running snaffle. (*Plate* 2.) They will see from that construction, that the purchase of it is greater than that of a common one. As its first point of *appui* is at the pommel of the saddle, lower than the rider's hand, they will also easily perceive, why they are good for horses, who have high light fore-hands, and why they are bad for such as have low and heavy ones. They are good for many horses, when used as a bridoon with a bridle, in cases of remarkably long, high fore-hands, and poking heads. On horses, whose heads and fore-hands are difficult to raise, a running snaffle, but not one fixed in the usual manner, is often very useful. The reins of it should be passed through an eye fixed on each side the head, pretty high up on the head-stall towards the ears, before they come into the rider's hand. (*Plate* 3.) When fixed at first to the rings on the head-stall, and coming through the eyes of the snaffle into the rider's hand, without being at all fixed to the saddle, they will

will often also be very useful. This lesson of the *Epaule en dedans*,[8] is a very touchstone in horsemanship, both for man and horse. Neither one nor the other can be dressed to any degree without a consummate knowledge of it; but it must not on any account be practised in the field in exercises, or evolutions: there the horses must always bend towards the side they are going, a thing (to the shame of the cavalry be it spoken) so rare to be seen. The *Epaule en dedans* reversed, is particularly advantageous to horses who are apt to throw themselves forward. By reversed, I mean when the shoulders are worked upon the outward larger circle, and the croup on the smaller circle next the center.

Horses well perfected in the *Epaule en dedans* may undertake, and soon learn any other lessons whatsoever. It ought, like all others, to be practised on all figures; circles, strait lines, squares, &c. and when on this last, which is an excellent lesson, (as also in every lesson, and on all figures, where there are corners and angles) care must be taken concerning the shoulders and croup, that, which ever of them is to enter the corner first, may go quite into it; and let that which goes in laſt, follow exactly the ſame ground. This rule can not be too much attended to. The croup, indeed, can never enter the corner firſt, except in working backwards.

---

*Of working in hand.*

WORKING in hand requires a certain degree of activity, a quick eye, and, like every thing elſe about horſes, good temper, and judgment. Though it can not be looked upon as a very difficult thing, I have ſeen few people ſucceed in it: none indeed, to any conſiderable degree, except Sir SIDNEY MEDOWS,[9] and the Cavaliere ROSSERMINI,[10] at Piſa, author of the *Cavallo Perfetto*. Begin by trotting, then galloping the horſe properly, bent inwards by a ſtrap tied from the ſide ring on the *caveſſon* to the ring on the pad. *(Plate* 4.) To the head-ſtall of the longe, a ſtrap and buckle under the throat is very uſeful to prevent the ſide part of it from chafing againſt the eye, which it is very apt to do, when the bending ſtrap is uſed, and drawn at all tight. Do this for a little while

while only at a time. If the horse leans on the strap which is tied to bend him, take off the *caveson*, and use in its stead one of the long strings which will be mentioned and explained a little further on, coming first from the ring on the pad, and from thence through the eye of the snaffle; *(Plate* 5.) and also, if the horse's head is low, through the ring on the head-stall, and from thence through the ring on the pad, *(Plate* 6.) into the hand of the person on foot, who must humour it, yielding and taking it up occasionally, which will prevent the horse's leaning, and make him light. *(Plate* 6.) The long string, thus used, will do very well alone, without the strap, when the horse is accustomed to bend, and to trot determined round the person who stands in the center, and holds the long string. After horses have been a little accustomed to be bent with a strap at the longe, they will very soon longe themselves, as it were; that is to say, that bent with the strap, they will go very well without any longe; and indeed, horses may be brought, with patience and gentleness, to work very well so on almost all lessons in hand. Next begin the *epaule en dedans*; and after that, the head to the wall, the croup to the wall, piaffing, backing, &c. on all figures, by degrees. I have observed, that most horses generally go the head to the wall more cordially at first, than they do the croup to the wall. Working in hand is, if I may be allowed the expreffion, a kind of driving. In explaining the method of working in hand, we will ufe the right all the way through. Two people on foot fhould be employed about it; one indeed may do, and well, if it is a handy perfon, but two are much better at firft: one of thefe people holds a long ftring, and in fome leffons two long ftrings, fixed, as fhall be prefently explained, and a *chambriere*,[11] ftanding at fome diftance from the horfe: the other perfon ftands near the horfe, holding the reins of the fnaffle, and a hand whip, to keep the horfe off from him, when neceffary. Girt a pad, with a crupper to it, upon the horfe: the pad muft have a large ring in the center upon the top of it, and, about four inches lower down on each fide, a fmaller one. On the top of the pad, a little forwarder than the great ring, there muft be a fmall ftrap, and buckle, which ferve to buckle in the fnaffle reins, and to prevent their floating about, and the horfe entangling his legs in them, in the longe. Horfes muft never be worked in hand with any thing in their mouths, but a large, thick, plain, running fnaffle: a bridle is too ticklifh, and would fpoil the horfe's mouth, unlefs it be in the hands of a very able mafter indeed; for, in working in hand, it is next to impoffible to be sufficiently gentle, and delicate with it. The eyes of the snaffle should be large, and on the head-stall, about the height of the horse's eye, should be fixed a ring on each side. The person with the *chambriere* holds a long string, about eighteen feet long, (so as to be out of the reach of the horse's heels) which must be smooth, of a proper thickness, and not stick, but run free. This string, in the *epaule en dedans*, (Plate 7.) to the right, is buckled to the right hand small ring on the pad, where the reins of the running snaffle are first fixed; from thence it passes through the right eye of the snaffle, and from that to the right hand small ring on the head-stall, and through the large ring on the top of the pad, into the hand of the person who holds the *chambriere*, and who, by means of this string, bends the horse to the right, and brings in his shoulder; following him on his right side, and tightening and loosening the string, as he finds it necessary. If the horse's fore-hand is high, and well placed, it will not be necessary to pass the string through the ring upon the head-stall: at the same time, another person standing near the horse, the snaffle reins separated, and the right one tied loose on the right side, leads him on with the left rein of the snaffle in his hand, walking near

his

his head, and taking care to keep the shoulders in their proper place; and not to take off from the bend to the right, which is occasioned by the string in the other person's hand; who will find it most convenient, when working on this lesson to the right, to hold the string in his right hand, and the *chambriere* in his left, and so *vice versâ*. These he must make use of, and keep himself more or less upon the flank, center, or rear of the horse, as he finds necessary. In the changes from right to left, in the *epaule en dedans*, the person nearest the horse must be quick in getting on the horse's left side; and the person with the *chambriere* must do the same; the former coming round by the horse's head before him, and the latter round by his croup behind him; and so *vice versâ* to the left. In the head, and in the croup, to the wall, both the men are already properly placed for the changes. In this lesson of the *epaule en dedans*, in hand, when a horse is very clumsy, heavy in hand, stiff, headstrong, vicious, or apt to strike with his fore feet, or to rear or kick out behind, a stick, or pole, is very useful; the stick, (about seven feet long) is fastened by a strap and buckle through the eye of the snaffle, where the reins pass: a man places himself, at a certain distance, on the side of the horse's head,

head, going before him over the ground to be worked upon, and holds the stick at arm's length, having tied it so, as to leave it room to play, as he draws it gently backwards and forwards to refresh and enliven the mouth. The other man holds a long rein, and the *chambriere*, as represented in *Plate* 7. Like the pillars, this lesson is excellent, or bad, according to the hands it is in. I have known a horse's jaw broke, and his tongue cut in two by it, and therefore it must be used in the most skilful and temperate manner, or not at all: it is useful in raising horse's heads; of those, particularly, who are apt to get their heads down, or to kick in piaffing on forwards, &c. Almost any lessons may be done by the help of this pole.

To work in hand, the head and the croup, to the wall, *(Plate* 8.) two strings fixed, as above described, (only that they must not come at all through the large ring on the pad, but from the small rings on the head-stall, immediately into the hand of the person who holds the *chambriere)* must be used, one on each side: one string, indeed, might do; the right one, in working to the right, and

so

so *vice versâ*: but two are much better, and often necessary, to help to keep the horse in a proper position. Passing the strings through the rings on the head-stall, is not necessary, when the horse carries his fore-hand high, and well; and when they do pass through them, great care must be taken, by a gentle use of them, that they do not gag the horse: these two strings must be buckled together, and meet in the hand of the person who holds the *chambriere*, and who is on the left side of the horse: the snaffle reins too must be joined, and the person near the horse, who holds them, must also be on the left side of him, and near his shoulder, holding the right rein of the snaffle the shortest, to bend him that way, (as does also the right string kept the tightest in the other person's hand) and making use also of the left rein, when necessary, to keep the horse in a proper position, and to guide him occasionally, as if he was upon him: and never so, as to take away from the bend. The lesson of the head, or croupe, to the wall, in hand, is often done better, when the man who follows, and holds the *chambriere*, has no long reins, or only one long rein, unless the horse is very aukward, refractory, or playful; for one of the long reins

reins is apt to get into the way of the man, who is nearer to the horfe. When only one long rein is ufed, it will be, of courfe, the right hand one, to the right, and fo *vice verfâ*. And indeed, in other leffons in hand, thefe long reins are no longer neceffary, when the horfe is grown handy; provided the man nearer to him has a feeling, fenfible, good hand, and perfectly knows what he is about. On the head or croup to the wall, in hand, it is a good way, at firft, to have a man, holding a long ftring buckled fimply to the eye of the fnaffle, go before the horfe, leading him, as it were, along the wall. Horfes will, with care and patience, not be very long before they work well in hand; though, indeed, never fo truly, or delicately, as under a good rider. Horfes worked well in hand look particularly well in coming up the middle, and backing there on the piaffer, as alfo in the piaffer, in one place, both bent, *(Plate 9.)* and ftraight, animated properly, and kept in a good pofition, their mouths being properly played with, and humoured. When horfes become free, and familiar with this method of working them in hand, it fhould be done by degrees on all paces, faft, and flow, but always without noise, hurry, or confusion. Nothing determines them better than working them in hand, when it is well done. As the want of great accuracy, and delicacy is, from the great numbers, in some measure unavoidable in military schools, it is not amiss to teach troop horses a little their lessons in hand, before the men do them on their backs. One of these strings may be used by the person who holds the *chambriere* on foot, when the horse is mounted; and it is a good method to do so, sometimes, on all lessons, and on all figures. This string fastened, as in the *epaule en dedans*, only that it goes immediately from the eye of the snaffle into the hand of the person on foot, who must stand in the center of the circle, helps the person upon the horse in the longe very much to bend him, as it does indeed in all other lessons. When the horse has a rider on him, only one string is necessary to be held by the person on foot. In the head to the wall, croup to the wall, piaffing, &c. &c. it must be shifted (for example, in the head to the wall, &c. &c. to the right) under the horse's jaw, from through the right eye of the snaffle, into the hand of the person on foot, who is on the left of the horse; for it need not pass through the small ring on the head-stall of the snaffle; the man upon the horse being the proper person to keep the horse's head up. It is sometimes expedient to pass the string over the horse's neck under the rider's hand, instead of under the horse's jaw. It must be fixed, in the first place, like a running snaffle, to the skirts of the saddle, from whence it goes, as above-mentioned, through the eye of the snaffle into the hand of the person on foot, after having passed under the horse's jaw. To piaffer too without any rider, on square, and all other figures, advancing gently, and well into the corners, is a very good lesson. One man must stand exactly before the horse, with his face to him, holding the two eyes of the snaffle, and keep the horse advancing gently, by going backwards himself. The man with the *chambriere* must stand behind the horse, and animate him, or not, as he finds necessary. Backing the horse so too sometimes is useful: that may also be done on all figures. The degree of vivacity, or dulness in the horse, must determine how the man with the *chambriere* is to act, and where he is to place himself, when the horse is backing. A horse when well taught may be worked, and it is then the best way, by a single man with long reins, and a *chambriere*, without any other person to assist. *(Plate* 10.) All airs in hand are to be worked so, whenever the animal is become supple and obedient.

Working in hand is very particularly useful in Military Equitation, because it spares the horse the fatigue of any weight upon him; and the want of a proper allowance of corn, to enable horses to go through the work with vigour, is a general army complaint, almost in all European services. When it is well done, it has a masterly, active appearance, and is always very useful in suppling and determining horses; but, past all doubt, a good rider mounted, who feels every motion of the horse, must act with more precision, delicacy, and exactness.

Great part of what has been said here, of working in hand, belongs properly to other chapters, but I was unwilling to divide the subject, and have therefore placed here what I had to mention about it.

CHAP.

## CHAP. IV.

*Of the head to the wall, and of the croup to the wall.*

THIS lesson should be practised immediately after that of the *epaule en dedans*, in order to place the horse properly the way he goes, &c. The difference between the head to the wall, and the croup to the wall, consists in this: in the former, the fore-parts are more remote from the center, and go over more ground; in the latter, the hinder-parts are more remote from the center, and consequently go over more ground: in both, as likewise in all other lessons, (those done in backing only excepted) the shoulders must go first. In riding-houses, the head to the wall is the easier lesson of the two, at first, the line to be worked upon being marked by the wall, which is not far from the horse's head. All lessons ought to be frequently varied, to prevent *routine*.

The motion of the legs in the lesson we are speaking of, to the right, is the same as that of the *epaule en dedans* to the left, and so *vice versâ*; but the head is always bent

and

and turned differently: in the *epaule en dedans*, the horse looks the contrary way to that which he goes; in this he looks the way he is going.

In the beginning, very little bend must be required; demanding too much at once would perplex the horse, and make him defend himself: it is to be augmented by degrees. If the horse absolutely refuses to obey, it is most probably a sign that either he or his rider has not been sufficiently prepared by previous lessons. It may happen, that weakness, or a hurt in some part of the body, or sometimes temper, though seldom, (in the horse I mean) may be the cause of the horse's defending himself: 'tis the rider's business to find out from whence the obstacle arises, and to remove it; and if he finds it to be from the first mentioned cause, the previous lessons must be resumed again for some time; if from the second, proper remedies must be applied; and if from the last cause, when all fair means that can be tried, have failed, proper corrections, with coolness and judgment, must be used.

In practising this lesson to the right, bend the horse to the right with the right rein, helping the left leg over the right,

right, (at the same time when the right leg is just come to the ground) with the left rein crossed towards the right, and keeping the right shoulder back with the right rein towards your body, in order to facilitate the left leg's crossing over the right; and so *vice versâ* to the left, each rein helping the other by their properly-mixed effects. In working to the right, the rider's left leg helps the hinder parts on to the right, and his right leg stops them, if they get too much so; and so *vice versâ* to the left; but neither ought to be used, 'till the hand, being employed, (as has before been explained) in a proper manner, has failed, or finds, that a greater force is necessary to bring what is required about, than it can effect alone; for the legs should not only be corresponding with the hand, but also subservient to it; and all unnecessary aids, as well as all force, ought always to be avoided as much as possible. In first beginning to teach this lesson, the croup must be but little constrained; as the horse grows more supple, engage it more by degrees.

In the execution of all lessons, the equilibre of the rider's body is of great use, ease and help to the horse: it ought

ought always to go with and accompany every motion of the animal; when to the right, to the right; and when to the left, to the left; if it does not, it is a very great hinderance to the horse's going.

This lesson is perpetually of service; for example, in all openings and closings of files: and though it be chiefly employed on straight lines, nevertheless it must be practised, advancing, retreating, turning, &c. as it may be of essential use almost in all cases whatever: it must be practised too in all paces, very fast as well as very slow, but of course gently at first; and changes also from one hand to the other must frequently be made on two pistes. 'Tis natural to imagine, that some horses, as well as some men, will be found more or less intelligent, active, vigorous, and supple, than others; and accordingly more or less is to be demanded and expected from them. This and all other lessons are to be performed with or without a longe, as may be found needful.

Upon all horses, in every lesson and action, it must be observed, that there is no horse but has his own peculiar *appui* or degree of bearing, and also a sensibility of mouth, as likewise a rate of his own, which it is absolutely necessary for the rider to discover and make himself acquainted with. A bad rider always takes off at least the delicacy of both, if not absolutely destroys it, which is generally the case. The horse will inform his rider when he has got his proper bearing in the mouth, by playing pleasantly and steadily with his bit, and by the spray about his chaps. A delicate and good hand will not only always preserve a light *appui*, or bearing in its sensibility, but also of a heavy one, whether naturally so or acquired, make a light one. The lighter this *appui* can be made, the better; but the rider's hand must correspond with it: if it does not, the more the horse is properly prepared, so much the worse for the rider. Instances of this inconvenience of the best of *appui*'s, when the rider is not equally taught with the horse, may be seen every day in some gentlemen, who try to get their horses bitted, as they call it, (which now and then, though very rarely, they get done to some degree) without being suitably prepared themselves for riding them: the consequence of which is, that they ride in danger of breaking their necks: 'till at length, after much hauling about, and by the joint insensibility and ignorance of themselves and their grooms, the poor animals gradually become mere senseless, unfeeling posts, and thereby grow, what they call, settled, and pleasant; that is to say, in reality, that they are grown as insensible as their riders, who, because they are void of feeling, and are not firm, must either hold by the bridle, or fall. One perpetually hears people say, they love a horse, who will let them bear a little on his mouth. Depend upon it, those people are not only ignorant, and unfeeling, but also very unfirm in their seat; for if they were not, they could not possibly find either use, or ease, in bearing a dead weight on their horses mouths. To help a horse every now and then, properly, is a very different, and a very useful thing. When the proper *appui* is found, and made of course as light as possible, it must not be kept dully fixed without any variation, but be played with; otherwise one equally continued tension of reins, though not a violent one, would render both the rider's hand, and the horse's mouth very dull. The slightest, and frequent giving, and taking is therefore necessary to keep both perfect.

Whatever pace or degree of quickness you work in, (be it ever so fast, or ever so slow) it must be cadenced; time is as necessary for an horseman, as for a musician.

Every soldier must be very well instructed in this lesson of the head and of the tail to the wall: scarce any manœuvre can be well performed without it. In closing and opening of files, it is almost every moment wanted. Few regimental riding-masters either practise it right, teach it right, or know it right, but act by force only: and make the horse look the wrong way. It is a great detriment to the service, that so few of the teachers are instructed on true and useful principles of horsemanship. This lesson of the head, or croup to the wall, &c. and all others, may be done on any pace; but, for the reasons given at the end of the sixth chapter, I shall give no very full instructions for them on a gallop here, as the nature of army riding hardly permits soldiers to be taught so far with exactness. If a horse is well taught on ever so slow a pace, he may, by degrees, without difficulty, be taught to do the same lesson with any degree of velocity. When he does it on a gallop, the rider must be quiet, and exact in the changes, and be then careful to ſtop the horſe's leg, with which he leads, juſt at the time when it is moſt forward, before it comes to the ground, by means of a ſlight tenſion of the rein on the ſame ſide, which will of courſe make the other leg go forward, and lead; and, that the horſe may change his hinder leg at the ſame time, which is abſolutely neceſſary, the rider muſt at the ſame time croſs over his hand, (to the left, for example, in changing from the left to the right) replacing it properly the moment the horſe has changed both before and behind, which muſt be done at the ſame time.

## CHAP. V.

*The Trot.*

THE three different kinds of trot, the extended, the supple, and the even, or equal, *(le determinè le deliè, & l'uni)* are explained so wonderfully masterly, and elegantly, in Monsieur BOURGELAT's *Nouveau Newcastle*,[12] that I can not omit giving here the chapter on trots of so truly admirable a master, for which I am obliged to Mr. BERENGER's translation of that excellent work.

" When a horse trots, his legs are in this position, two in the air, and two upon the ground, at the same time crosswise; that is to say, the near foot before, and the off foot behind are off the ground, and the other two upon it, and so alternately of the other two. This action of his legs is the same as when he walks, except that in the trot his motions are more quick."[13] All writers, both ancient and modern, have constantly asserted the trot to be the foundation of every lesson you can teach a horse: there are

are none, likewife, who have not thought proper to give general rules upon this fubject, but none have been exact enough to defcend into a detail of particular rules, and to diftinguifh fuch cafes as are different, and admit of exceptions, though fuch often are found from the different make and tempers of horfes, as they happen to be more or lefs fuited to what they are deftined; fo that, by following their general maxims, many horfes have been fpoiled, and made heavy and aukward, inftead of becoming fupple and active, and as much mifchief has been occafioned by adopting their principles, although juft, as if they had been fuggefted by ignorance itfelf. Three qualities are effentially neceffary to make the trot ufeful. It ought to be extended, fupple, and even, or equal. Thefe three qualities are related to, and mutually depend upon each other; in effect, you cannot pafs to the fupple trot, without having firft worked upon the extended trot: and you can never arrive at the even and equal trot, without having firft practifed the fupple. I mean by the extended, that trot, in which the horfe trots out without retaining himfelf, being quite ftrait, and going directly forwards; this confequently is the kind of trot with which you muft
begin;

begin; for before any thing else should be thought of, the horse should be taught to embrace, and cover his ground readily, and without fear. The trot however may be extended without being supple, for the horse may go directly forward, and yet not have that ease, and suppleness of limb, which distinguishes, and characterises the supple. I define the supple trot to be that, in which the horse at every motion that he makes, bends and plays all his joints, that is to say, those of his shoulders, his knees, and feet, which no colts or raw horses can execute, who have not had their limbs suppled by exercise, and who generally trot with a surprizing stiffness, and aukwardness, without the least spring or play in their joints. The even or equal trot, is that wherein the horse makes all his limbs and joints move so equally, and exactly, that his legs never cover more ground one than the other, nor at one time more than another. To do this, the horse must of necessity unite and collect all his strength, and, if I may be allowed the expression, distribute it equally through all his joints. To go from the extended trot to the supple, you must gently, and by degrees hold in your horse, and when by exercise he has attained sufficient ease and suppleness to manage his limbs readily, you muſt inſenſibly hold him in ſtill more and more, and by degrees you will lead him to the equal trot. The trot is the firſt exerciſe to which a horſe is put; this is a neceſſary leſſon, but, if given unſkilfully, it loſes its end, and even does harm. Horſes of a hot, and fretful temper, have generally too great a diſpoſition to the extended trot; never abandon theſe horſes to their will, hold them in, pacify them, moderate their motions by retaining them judiciouſly; their limbs will grow ſupple, and they will acquire at the ſame time that union and equality which is ſo eſſentially neceſſary. If you have a horſe that is heavy, conſider if this heavineſs, or ſtiffneſs of his ſhoulders, or legs, is owing to a want of ſtrength, or of ſuppleneſs; whether it proceeds from his having been exerciſed unſkilfully, too much, or too little. If he is heavy, becauſe the motions of his legs and ſhoulders are naturally cold, and ſluggiſh, though at the ſame time his limbs are good, and his ſtrength is only confined, and ſhut up, if I may ſo ſay, a moderate, but continual exerciſe of the trot will open and ſupple his joints, and render the action of his ſhoulders and legs more free, and bold; hold him in the hand, and ſupport him

in

in his trot, but take care to do it so as not to check, or slacken his pace; aid him, and drive him forward while you support him; remember at the same time, that if he is loaded with a great head, the continuation of the trot will make his *appui* hard and dull, because he will by this means abandon himself still more, and weigh upon the hand. All horses that are inclined to be *ramingue*,[14] that is to say, to retain themselves, and to resist by so doing, should be kept to the extended trot. Every horse, who has a tendency to be *ramingue*, is naturally disposed to unite himself, and collect all his strength; your only way with such horses is to force them forward; in the instant that he obeys, and goes freely on, retain him a little, yield your hand immediately after, and you will find soon that the horse of himself will bend his joints, and go united and equally. A horse of a sluggish and cold disposition, which has nevertheless strength and bottom, should likewise be put to the extended trot. As he grows animated, and begins to go free, keep him together by little and little, in order to lead him insensibly to the supple trot: but if while you keep him together, you perceive that he slackens his action, and retains himself, give him the aids briskly, and push him forward, keeping him nevertheless gently in hand; by this means he will be taught to trot freely, and equally at the same time. If a horse of a cold, and sluggish temper, is weak in his legs, and reins, you must manage him cautiously in working him in the trot, otherwise you will enervate, and spoil him. Besides, in order to make the most of a horse who is not strong, endeavour to give him wind, by working him slowly, and at intervals, and by encreasing the vigour of his exercise by degrees; for you must remember, that you ought always to dismiss your horse before he is spent, and overcome by fatigue; never push your lessons too far, in hopes of suppling your horse's limbs by the trot, instead of this you will falsify, and harden his *appui*, which is a case that happens but too frequently. Farther, it is of importance to remark, that you ought at no time, neither in the extended, supple, or equal trot, to confine your horse in the hand, in expectation of raising him, and fixing his head in a proper place. If his *appui* be full in hand, and the action of his trot should be checked, and restrained by the power of the bridle, his bars would very soon grow callous, and his mouth be hardened,

and

and dead; if, on the contrary, he has a fine, and sensible mouth, this very restraint would offend, and make him uneasy; you must endeavour then, as has already been said, to give him by degrees, and insensibly, the true and just *appui*, to place his head, and form his mouth by stops, and half-stops, by sometimes moderating and restraining him, with a gentle, and light hand, and yielding it to him immediately again, and by sometimes letting him trot without feeling the bridle at all. There is a difference between horses who are heavy in the hand, and such as endeavour to force it: the first sort lean, and throw all their weight upon the hand, either as they happen to be weak, or too heavy, and clumsy in their foreparts, or from having their mouths too fleshy and gross, and consequently dull and insensible: the second pull against the hand, because their bars are hard, lean, and generally round: the first may be brought to go equal, and upon their haunches, by means of the trot, and slow gallop; and the other may be made light and active by art, and by settling them well in their trot, which will also give them strength and vigour. Horses of the first sort are generally sluggish; the other kind are, for the most part, impatient, and difobedient, and upon that very account more dangerous, and incorrigible. The only proof, or rather the moſt certain ſign of your horſe's trotting well, is, that when he is in his trot, and you begin to preſs him a little, he offers to gallop. After having trotted your horſe ſufficiently upon a ſtrait line, or directly forwards, work him upon circles, but before you put him to this, walk him gently round the circle, that he may apprehend and know the ground he is to go over. This being done, work him in the trot. A horſe that is loaded before, and heavily made, will find more pains and difficulty in uniting his ſtrength, in order to be able to turn, than in going ſtrait forward. The action of turning tries the ſtrength of his reins, and employs his memory and attention; therefore let one part of your leſſons be to trot them ſtrait forward: finiſh them in the ſame manner, obſerving that the intervals between the ſtops (which you ſhould make very often) be long, or ſhort, as you judge neceſſary. I ſay, you ſhould make frequent ſtops, for they often ſerve as a correction to horſes that abandon themſelves, force the hand, or bear too much upon it in their trot. There are ſome horſes who are ſupple in

their

their shoulders, but who neverthelefs abandon themfelves, this fault is occafioned by the rider's having often held his bridle hand too tight in working them upon large circles; to remedy this, trot them upon one line or tread, and very large; ftop them often, keeping back your body and outward leg, in order to make them bend and play their haunches. The principal effects then of the trot are to make a horfe light, and active, and to give him a juft *appui*. In reality, in this action he is always fupported on one fide by one of his fore legs, and on the other by one of his hind legs: now the fore and hind parts being equally fupported crofswife, the rider cannot fail of fuppling, and loofening his limbs, and fixing his head; but if the trot difpofes, and prepares the fpirits and motions of a finewy and active horfe for the jufteft leffons, if it calls out and unfolds the powers, and ftrength of the animal, which before were buried, and fhut up, if I may ufe the expreffion, in the ftiffnefs of his joints and limbs; if this firft exercife, to which you put your horfe, is the foundation of all the different airs, and maneges, it ought to be given in proportion to the ftrength and vigour of the horfe. To judge of this, you muft go farther than mere out-

ward

ward appearances. A horse may be but weak in the reins, and yet execute some air, and accompany it with vigour, as long as his strength is united and entire; but if he becomes disunited, by having been worked beyond his ability in the trot, he will then faulter in his air, and perform it without vigour or grace. There are also some horses who are very strong in the loins, but who are weak in their limbs; these are apt to retain themselves, they bend, and sink in their trot, and go as if they were afraid of hurting their shoulders, their legs or feet. This irresolution proceeds only from a natural sense they have of their weakness. This kind of horses should not be too much exercised in the trot, nor have sharp correction; their shoulders, legs, or hocks, would be weakened and injured; so that learning in a little time to hang back, and abandon themselves on the *appui*, they would never be able to furnish any air with vigour, and justness. Let every lesson then be well weighed; the only method by which success can be insured, is the discretion you shall use in giving them in proportion to the strength of the horse, and from your sagacity in deciding upon what air or manege is most proper for him, to which you must be directed by observing which seems most suited to his inclination and capacity.

CHAP.

## CHAP. VI.

*The method of reining back---and of moving forwards immediately after---of piaffing---of pillars, &c.---of moving pillars, &c.*

SOMETHING having already been said, in the chapter of suppling, &c. upon the subject of reining back, there will not be occasion to dwell much upon it here, as the reader may have recourse to that chapter. Horses, particularly such as are never put in the pillars, nor taught to piaffe, should be reined back a good deal, sometimes slow, sometimes fast, and always without confusion, both in hand, and when rode. Never finish your work by reining back, especially with horses that have any disposition towards retaining themselves; but always move them forwards, and a little upon the haunches also after it, before you dismount; unless they retain themselves very much indeed, in which case nothing at all must be demanded from the haunches, but, quite the contrary, they must immediately be trotted hard out. This lesson of reining back,

back, and piaffing, is excellent to conclude with, and puts a horse well and properly on the haunches: the head and fore-parts must be kept high, and free, for any confinement there destroys action. To bend the horses sometimes in doing it, is a good lesson. It may be done, according as horses are more or less suppled, either going forwards, backing, or in the same place: if 'tis done well advancing, or at most, on the same spot, it is full sufficient for a soldier's horse: for to piaffe in backing, is rather too much to be expected in the hurry, which cannot but attend such numbers both of men and horses, as must be taught together in regiments. This lesson must never be attempted at all, 'till horses are very well suppled, and somewhat accustomed to be put together; otherwise it will have very bad consequences, and create **restiveness**: infallibly so, if not practised with the utmost exactness and delicacy; and principally with horses, that have the least tendency to retain, or to defend themselves. If they refuse to back, and stand motionless, the rider's legs must be approached with the greatest gentleness to the horse's sides; at the same time as the hand is acting on the reins to solicit the horse's backing. This seldom fails of procuring

curing the desired effect, by raising one of the horse's fore legs, which being in the air, has no weight upon it, and is consequently very easily brought backwards by a small degree of tension in the reins. When this lesson of piaffing is well performed, it is very noble, and useful, and has a pleasing air; it is an excellent one to begin teaching scholars with. In regiments, at their first being raised, when all horses are brought in young and raw, there can of course be no horses ready prepared in it for this purpose; but a litle time and diligence remedies this inconvenience.

The lesson, we are speaking of, is particularly serviceable in the pillars, for placing scholars well at first. Very few regimental riding-houses have pillars, and I must say, that it is fortunate they have not; for though, when properly made use of with skill, they are one of the greatest and best discoveries in horsemanship, they must be allowed to be very dangerous and pernicious, when they are not under the direction of a very knowing person. Upon the whole, however highly I approve of pillars, I would on no account admit of any, unless constantly under the eye and attention of a very intelligent teacher; which is a thing so difficult to be found in regiments, that I think pillars are better banished from amongst them, and therefore shall say no more here of what I esteem nevertheless so much. As for the single pillar, used in the manner it formerly was, it is a very useless and ridiculous thing; and being, as I hope and believe, universally laid aside, I think it not worth making further mention of here. Moving pillars are exempt from those inconveniences which attend fixed ones, and I therefore recommend them for army riding. By moving pillars, I understand a horse held by a rein on each side, by a man on each side of him: another person with a *chambriere* follows, animates, or sooths him, as he finds necessary, and makes him piaffe backwards, or forwards, with, or without long reins, as is found expedient. When the long reins, or strings are used, or rather the long string or rein, (for one is generally sufficient) it must be fixed on the side the horse is to be bent: this string is fixed to the saddle, and goes through the eye of the snaffle, and also through a ring on the head-stall, if the horse is apt to get his head low: one man, besides the one who holds the *chambriere*, is sufficient in this case: the horse is bent to the right, or left, or kept wholly strait. This method

thod is particularly useful for horses whose action of their hinder legs is confined, and wants liberty: the same rule will hold good for all horses so circumstanced in all they do; for they should always be worked boldly out on large scales, and never confined to small figures. A horse looks remarkably well in this attitude, if those who hold him have light hands, and keep his head high: they should each of them have a switch, to help them to keep the horse straight, in case of necessity. This lesson may be very well done by one man alone, with long reins *(as in Plate* 10.)

It would scarce be possible (neither is it indeed necessary) to teach the more refined and difficult parts of horsemanship to all the different kinds, and dispositions, both of men and horses, which are in all regiments; or to find the time and attention requisite for it to such numbers; but I yet hope some proper institution will be formed, to make good riding-masters, farriers, sadlers, and gun-smiths, and every thing else necessary for the army, upon a good, and proper footing: they are absolutely necessary, and should be properly and equally divided through the regiment, in the squadrons and troops. There should be one riding-master in chief, with a sufficient number of under ones under him, and formed by him; he should inspect the work of the others very frequently, and give lessons by turns to the whole regiment, going about from one quarter to another, if the regiment is separated: he should break too the officers horses, or rather teach them to do it themselves, who, I am sorry to say it, stand at present, in general, in the greatest need of instructions,---no people more: they should, therefore, and for the sake of creating emulation too in the men by their example, always attend the riding-master regularly two or three times a week, at least. I must urge the necessity of forming by reading, and serious study, as well as by much constant practice, proper riding-masters for the army; though I am thoroughly apprized, as the celebrated Mr. BOURGELAT observes, that an ill-founded prejudice partially directs the judgment of the greater part of those people, who call themselves conoisseurs. I know full well that they suppose that practice alone can insure perfection, and that in their arguments in favour of this their deplorable system, they reject with scorn all books, and authors: but Equitation is confessedly a science; every science is founded upon principles, and theory

must

must indispensably be necessary, because what is truly just and beautiful can not depend upon chance. What indeed is to be expected from a man, who has no other guide than a long continued practice, and who must of necessity labour under very great uncertainties! Incapable of accounting rationally for what he does, it must be impossible for him to enlighten me, or communicate to me the knowledge which he fancies himself possessed of. How then can I look upon such a man as a master? On the other hand, what advantages may I not obtain from the instructions of a person, whom theory enables to comprehend and feel the effects of his slightest operations, and who can explain to me such principles, as an age of constant practice only could never put me into a way of acquiring? Equitation does, to be sure, require also a constant, and an assiduous exercise. Habit, and continual practice will go a great way in all exercises, which depend on the mechanism of the body, but, unless this mechanism is properly fixed, and supported on the solid basis of theory, errors will be the inevitable consequence. In working a horse, a principal object should be to exercise the genius, and memory of the animal, as well as his body. You

should

should endeavour to discover his natural inclination, and to get a thorough knowledge of his abilities, in order to take advantage in future of that knowledge. Without the help of lights derived from just principles, it is morally impossible that a horseman should make use of his reason upon all occasions, or be able to find out, with care and attention, whatever may conduct him to the end and object of his hopes, desires, and undertakings; because, in few words, there is an absolute necessity of some method for improving the natural disposition of the animal, which is in some cases defective and intractable. The consequences of the false, and prejudicial system, which I am opposing, justify my assertions. The knowledge of a horse is vulgarly thought so familiar, and the means of dressing[15] him so general, and so common, that you can hardly meet with a man, who does not flatter himself, that he has succeeded in both points; and while masters, who sacrifice every hour of their life to attain knowledge, still find themselves immerged in darkness and obscurity, men the most uninformed imagine, that they have attained the summit of perfection, and in consequence thereof suppress the least inclination of learning even the first elements.

A blind, and boundless presumption is the characteristic of ignorance; the fruits of long study, and application amount to a discovery of innumerable fresh difficulties, at the sight of which a diligent many very far from over-rating his own merit, redoubles his efforts in pursuit of further knowledge.

# A METHOD OF

## CHAP. VII.

*The method of teaching horses to stand fire, noises, alarms, sights, &c.---of preventing their lying down in the water---to stand quiet to be shot off from---to go over rough and bad ground ---to leap hedges, gates, ditches, &c. standing and flying--- to disregard dead horses---to swim, &c.*

IN order to make horses stand fire, the sound of drums, and all sorts of different noises, you must use them to it by degrees in the stable at feeding-time; and instead of being frightened at it, they will soon come to like it, as a signal for eating.

With regard to such horses as are afraid of burning objects, begin by keeping them still at a certain distance from some lighted straw: caress the horse, and in proportion as his fright diminishes, approach gradually the burning straw very gently, and increase the size of it. By this means he will very quickly be brought to be so familiar with it, as to walk undaunted even through it. The same

ſame method and gentleneſs muſt be obſerved alſo, in regard to glittering arms, colours, ſtandards, &c.

As to horſes that are apt to lie down in the water, if animating them, and attacking them vigorouſly, ſhould fail of the deſired effect, (which ſeldom is the caſe) then break a ſtraw-bottle full of water upon their heads, the moment they begin to lie down, and let the water run into their ears, which is a thing they apprehend very much, and which will in all probability ſoon cure them of the trick.

All troop-horſes muſt be taught to ſtand quiet and ſtill when they are ſhot off from, to ſtop the moment you preſent, and not to move after firing, 'till they are required to do it: this leſſon ought eſpecially to be obſerved in light troops, and it ſhould never be neglected in any kind of cavalry whatſoever: in ſhort, the horſes muſt be taught to be ſo cool and undiſturbed, as to ſuffer the riders to act upon them with the ſame freedom, as if they were on foot. Patience, coolneſs, and temper, are the only means requiſite for accompliſhing this end.

The rider, when he fires, muſt be very attentive not to throw himſelf forwards too much, or otherwiſe *derange* himſelf in his ſeat. Begin by walking the horſe gently, then ſtop and keep him from ſtirring for ſome time, ſo as to accuſtom him by degrees not to have the leaſt idea of moving without orders: if he does, back him; and when you ſtop him, and he is quite ſtill, leave the reins quite looſe, and careſs him.

To uſe a horſe to fire-arms, firſt put a piſtol or carbine in the manger with his feed; then uſe him to the ſound of the lock and the pan; after which, when you are upon him, ſhew the piece to him, preſenting it forwards, ſometimes on one ſide, ſometimes on the other: when he is thus far reconciled, proceed to flaſh in the pan; after which, put a ſmall charge into the piece, and ſo continue augmenting it by degrees to the quantity which is commonly uſed: if he ſeems uneaſy, walk him forwards a few ſteps ſlowly, and then ſtop, back, move forwards, then ſtop again, and careſs him. Great care muſt be taken not to burn, or ſinge the horſe any where in firing; he would remember it, and be very ſhy, for a long time. Horſes are

are also often disquieted and unsteady at the clash and glittering of arms, at the drawing and returning of swords, all which they must be familiarized to by little and little, by frequency and gentleness.

In going over rough and bad ground, the men must keep their hands high, and their bodies back.

It is very expedient for all cavalry, in general, but particularly for light cavalry,[16] that their horses should be very ready and expert in leaping over ditches, hedges, gates, &c. not only singly but in squadrons,[17] and lines.[18] The leaps, of whatever sort they are, which the horses are brought to in the beginning, ought to be very small ones, and as the horse improves in his leaping, be augmented by degrees; for if the leaps were encreased considerably at once, the horse would blunder, grow fearful, and contract an aukward way of leaping with hurry, and confusion. The riders must keep their bodies back, raise their hand a little in order to help the fore-parts of the horse up, and be very attentive to their equilibre, without raising themselves up in the saddle, or moving their arms. The surest way to prevent people, in leaping over any thing, from raising up their arms and elbows, (which is an unfirm, and ungraceful motion) is to make them put a hand whip, or switch, under each arm, and not let them drop. 'Tis best to begin at a low bar covered with furze,[19] (*Plate* 15 .☉.) which pricking the horse's legs, if he does not raise himself sufficiently, prevents his contracting a sluggish and dangerous habit of touching, as he goes over, which any thing yielding, and not pricking, would give him a custom of doing. Many horses, in learning to leap, are apt to come too near, and in a manner with their feet under the bar. The best way to prevent their doing so, is to place under the bar two planks of the breadth of the pillars on which the leaping bar is fixed: these planks must meet and join at top under the bar, about two feet high from the ground, (*Plate* 15 .+.) and project at bottom upon the ground, about two feet; they must be strongly framed, that the horse may not break them, by touching them with his feet. The bar should be placed so as to run round, when touched. Let the ditches and hedges, &c. you first bring horses to, be inconsiderable, and in this, as in every thing else, let the increase be made by degrees. Accustom them to come up gently to every thing, which they

are

# BREAKING HORSES, &c.

are to leap over, and to stand coolly at it for some time; and then to raise themselves gently up, and go clear over it, without either sloth or hurry. When they leap well standing, *(Plate* 11 *and* 13.) then use them to walk gently up to the leap, and to go over it without first halting at it; and after that practice is familiar to them, repeat the like in a gentle trot, and so by degrees faster and faster, 'till at length it is as familiar to them to leap flying on a full gallop, *(Plate* 12 *and* 14.) as any other way; all which is to be acquired with great facility by calm and soft means, without any hurry.

As horses are naturally apt to be frightened at the sight and smell of dead horses, numbers of which are every moment met with on service, (especially at the latter end of the year, when the roads are bad, and the poor animals, too often treated and driven cruelly, go a great way from camp for forage) it is adviseable to habituate them to walk over, and leap over carcases of dead horses; and as they are particularly terrified at this sight, the greater gentleness ought consequently to be used in breaking them of their dread of it.

Horses should also be accustomed to swim, which often may be necessary upon service; and if the men and horses both are not used to it, both may be frequently liable to perish in the water. A very small portion of strength is sufficient to guide a horse, any where indeed, but particularly in the water, where they must be permitted to have their heads, and be as little constrained as possible in any shape. In crossing rivers, the horse's head should be kept against the current, more or less, according to the situation of the place, higher up, or lower down, purposed to land at, and the degree of rapidity of the water. In going down the stream, the straighter the horse is the better. The rider had always better quit his stirrups on these occasions, for fear of accidents, and his getting entangled in them. A horse is turned difficultly in the water; it must be done very gently and carefully. For partizans,[20] and all who go chiefly on reconnoitring duty, horses should be chosen, who are not apt to neigh: the Numidians[21] preferred mares to horses, for surprizes on the enemy, because, being less apt to neigh, they were less likely to be discovered. Those of the whole army should be taught to be obedient to the voice, and to carry double.

ble. Reins may be cut in battle; and in crossing waters, and upon forced marches, it may sometimes be necessary to take the infantry *(en croupe)* up behind. The ancient Lybians directed their horses in battle by the voice; and the same custom prevails amongst them to this day, for the modern Africans do the same.

The heavy cavalry may possibly object to having their large horses taught all these several exercises; but though they are not, nor can indeed be expected to perform all, with the same activity and velocity, as light troops do, yet 'tis absolutely necessary, that they should be taught them all; for 'tis a melancholy consideration, that any trifling obstacle should prevent so useful and powerful a body from acting. I cannot take upon me to say, whether it was always so in former times, or not: the ancients, I believe, understood horsemanship more than we are aware of: there is a great deal of good sense in XENOPHON's[22] method of forming horses for war; after him, horsemanship was buried for ages, or rather brutalised, which is still too much the case.

CHAP.

## CHAP. VIII.

*The method of curing restivenesses, vices, defences, starting, and stumbling, &c.*

BEFORE any mention is made of the different kinds of restivenesses, vices, and defences, &c. it is not amiss to observe, that a horse's being good or ill-natured greatly depends on the temper of the person, that is put about him, especially at first; and consequently one cannot be too careful and watchful in this point.

Whenever a horse makes resistance, one ought, before a remedy or correction is thought of, to examine very minutely all the tackle about him, if any thing hurts or tickles him, whether he has any natural or accidental weakness, or in short any the least impediment in any part. For want of this precaution, and previous inspection, many fatal, and often irreparable disasters happen: the poor dumb animal is frequently accused falsely of being restive and vicious; is used ill without reason, and being forced into

despair,

despair, is, in a manner, obliged to act accordingly, be his temper and inclination ever so well disposed. It must never be forgot, that it is necessary to work on the minds of horses, at first by slow motions which give them time to reflect. By degrees every thing may be done most rapidly with ease and very well. Such is in general, unless spoilt by us, the good temper, docility, and obedience of a horse, that almost any thing may be done with him by good-nature, and science. Even the domestic, worthy, friendly dog is not more susceptible of education.

A horse that is vicious and also so weak, that there are no hopes of his growing stronger, is a most deplorable beast, and not worth any one's care or trouble: 'tis very seldom, (I was near saying, never) the case, that a horse is really, and by nature vicious; but if such be found, chastisements will become necessary sometimes, but they must then be always made use of with the greatest judgment, and temper. The propriety of aids is to foresee, and prevent faults. The propriety of chastisements is to correct them.

Correction, according as you use it, throws a horse into more or less violent action, which, if he be weak, he cannot support: but a vicious strong horse is to be considered in a very different light, being able both to undergo and consequently to profit by all lessons; and is, in every respect, far preferable to the best-natured weak one upon earth. Patience and science are never-failing means to reclaim a wicked horse: in whatsoever manner he defends himself, bring him back frequently with gentleness, but with firmness too, to the lesson which he seems most averse to, Horses are by degrees made obedient through the hope of recompence and the fear of punishment: how to mix these two motives judiciously together is a very difficult matter, not easy to be prescribed; it requires much thought and practice; and not only a good head, but a good heart likewise. The coolest, and best-natured rider, *cæteris paribus*,[23] will always succeed best. By a dextrous use of the incitements above-mentioned you will gradually bring the horse to temper and obedience; mere force and want of skill and of coolness would only tend to confirm him in bad tricks. If he be impatient or choleric, never strike him, unless he absolutely refuses to go forwards,

which

which you must resolutely oblige him to do, and which will be of itself a correction, by preventing his having time to meditate, and put in execution any defence by retaining himself. Resistance in horses, you must consider, is sometimes a mark of strength and vigour, and proceeds from spirits, as well as sometimes from vice and weakness. Weakness frequently drives horses into viciousness, when any thing, wherein strength is necessary, is demanded from them; nay, it inevitably must: great care therefore should always be taken to distinguish from which of these two causes, that are evidently so different, the defence arises, before any remedy or punishment is thought of. It may sometimes be a bad sign, when horses do not at all defend themselves, and proceed from a sluggish disposition, a want of spirit, and of a proper sensibility. Whenever one is so fortunate as to meet with a horse of just the right spirit, activity, delicacy of feeling, with strength, and good-nature, he cannot be cherished too much; for such a one is a rare and inestimable jewel, and if properly treated, will, in a manner, do every thing of himself. Horses are oftener spoilt by having too much done to them, and by attempts to dress them in too great a hurry, than by any other treatment.

If after a horse has been well suppled, and there are no impediments, either natural or accidental, if he still persists to defend himself, chastisements then become necessary: but whenever this is the case, they must not be frequent, but always firm, though always as little violent, as possible: for they are both dangerous and very prejudicial, when frequently or slightly played with; and still more so, when used too violently. When a rider quarrels with his horse, he is generally the dupe of his passion, and the fray commonly ends to his disadvantage. Whenever you see a man beating any animal, you will almost always find, that the man is in the wrong, and the animal in the right.

'Tis impossible in general, to be too circumspect in lessons of all kinds, in aids, chastisements or caresses; for as the great Duke of Newcastle[24] observes, if any man was in the form of a horse, he could not invent with more art than some horses do, schemes to oppose what is required of him. Some have quicker parts, and more cunning, than others. Many will imperceptibly gain a little every day on their rider. Various in short are their dispositions, and capacities. It is the rider's business to find out their different qualities,

and

and to make them sensible how much he loves them, and desires to be loved by them, but at the same time, that he does not fear them, and will be master. A good natured clever man may with the greatest ease teach a horse any thing; more tricks even of all kinds, than dogs are seen to perform at fairs. Plunging is a very common defence among restive and vicious horses: if they do it in the same place or backing, they must by the rider's legs, and spurs too sometimes firmly applied, be obliged to go forwards, and their heads kept up high. But if they do it flying forwards, keep them back, ride them gently and very slow for a good while together, and back them gently every now and then. Of all bad tempers and qualities in horses, those, which are occasioned by harsh treatment and ignorant riders, which are very common, are the worst.

Rearing is a bad vice, and in weak horses especially, a very dangerous one. Whilst the horse is up, the rider must yield his hand, and when the horse is descending he must vigorously determine him forwards by approaching his legs to the horse's sides: if this be done at any other time, but whilst the horse is coming down, it may add a spring to his rearing,

rearing and make him fall backwards. With a good hand on them, horses seldom persist in this vice; for they are themselves naturally much afraid of falling backwards. If this method, which I have mentioned, fails, (which it scarcely ever will) you must make the horse kick up behind, by getting somebody on foot to strike him behind with a whip; or, if that will not effect it, by pricking him with a goad.

Starting often proceeds from a defect in the sight, which therefore must be carefully looked into. Whatever the horse is afraid of, bring him up to it gently; if you caress him every step he advances, he will go quite up to it by degrees, and soon grow familiar with all sorts of objects. Nothing but great gentleness can correct this fault: for if you inflict punishment, the apprehension of chastisement becomes prevalent, and causes more starting, than the fear of the object. If you let him go by the object, without bringing him up to it, you increase the fault and confirm him in his fear: the consequence of which is, he takes his rider perhaps a quite contrary way from what he was going, becomes his master, and puts himself and the person upon him, every moment in great danger. I have so often heard people maintain, some, that blows are necessary to cure this

this evil; and others, that horses should be suffered to have their own way in it, that I could not help saying a few words upon this subject, (though it speaks for itself) to convince those, who, as my ingenious friend Mr. BOURGELAT says, *argumentent de ces systemes deplorables.*[25]

Quarrelling with horses, plaguing, or beating them, as one often sees done, not only spoils both their tempers, and their paces, but it teaches them to trip, stumble, fall, start, run away, and to be unsteady and vicious, &c. whilst gentleness and coolness would very soon bring them to go through, or over any bad place whatsoever, with ease, good-humour and safety. Beat a horse for a trip, or such a kind of thing, and he will soon do it again through fear and hurry. Such failures sometimes proceed from weakness. In that case, proper food, and gentle exercise, by restoring the animal to health, and vigour, will cure him of them. If they come from inattention, or from the badness of his paces, he must have a good rider to render him attentive, and mend his movements. All other remedies will prove fruitless, but these will not, unless some natural defects, or acquired hurts, such as lameness, or bad weakening distempers interfere. Many

Many troop horses, and particularly old ones, often do not chuse to leave their companions. They should therefore be used early, and frequently to leave their ranks singly.

With such horses, as are to a very great degree fearful of any objects, make a quiet horse, by going before them, gradually entice them to approach nearer and nearer to the thing they are afraid of. If the horse, thus alarmed, be undisciplined and headstrong, he will probably run away with his rider; and if so, his head must be kept up high, and the snaffle sawed backwards and forwards from right to left, taking up and yielding the reins of it, as also the reins of the bit: but this latter must not be sawed backwards and forwards, like the snaffle, but only taken up, and yielded properly. No man ever yet did, or ever will stop a horse, or gain any one point over him by main force, or violence, or by pulling a dead weight against him.

Upon horses, who have a trick of turning short about suddenly, to the right for example, seperate the reins, taking one in each hand: leave the right one quite loose, and pull the left one, stretching out your hand from the horse to the left, and forwards. If the horse still resists, use your left leg, and spur; and so *vice versâ*, 'till he turns to the left.

CHAP.

## CHAP IX.

*Several remarks and hints on shoeing, feeding, management of horses, &c. &c.*

I Do not by any means intend to enter here largely on the many systems of shoeing; it would enlarge this treatise too much, and extend the object of it beyond the bounds I have prescribed to it, and to myself: as feet differ, so should shoes accordingly, but as it happens unfortunately for us, that the farriers belonging to the army, for want of proper education, due inspection, and encouragement, are void of all real skill, and knowledge in their profession, and have minds, in short, quite uncultivated, it is absolutely necessary to lay down only such rules, as are plain, general and invariable, and the strictest discipline must be enforced to make them all observed and followed most religiously. I do not however despair of seeing in time some intelligent farriers properly instructed; and when such are formed, and not 'till then, the number of them in regiments should be increased: It would even be much better to have none at all, 'till such a reformation is brought about. One man cannot properly shoe more than forty horses; at present we have only one to a troop of fifty-five, in time of war, besides bat-horses, and all others belonging to officers, sutlers, carriages, servants, &c. There should also be one forge-cart[26] at least appropriated to each squadron, and a third for the latter-mentioned purposes: but they must not be like our present ones, which are made so heavy and with such low wheels, that they employ a great number of horses, ruin most of them, and after all, seldom get up to their respective regiments in right time, even in good roads, and never in bad ones. And I may say, that 'tis lucky they do not, for upon experience one finds fewer horses lame, during the absence of farriers, than when they are present. They should be built upon two wheels only, and those very high: The cart must be covered, and have partitions in it for the forge, bellows, tools, charcoal, &c. All these things must be so contrived, as to be easily taken out of the cart, and worked on the ground. This sort of forge-cart never sticks, and is always able to keep up with the regiments on any marches: it requires but few horses, and spoils none. I have one for my own use, made by the Hanoverian train,[26]

which

which is drawn easily by two horses. For regiments, the carts must be somewhat larger, and more substantial, and would require three horses. I doubt not, but an English workman would improve upon them, as to strength and lightness, as well as convenience; tho' the cart I have, is very well constructed, and answers well every necessary purpose.

Physic and a butteris[28] in well-informed hands would not be fatal; but in the manner we are now provided with farriers, they must be quite banished. Whoever lets his farrier, groom, or coachman, ever even mention any thing more than water-gruel, a clyster,[29] or a little bleeding, and that too very seldom; or pretend to talk of the nature of feet, of the seat of lamenesses, sicknesses, or their cures, may be certain to find himself very shortly, and very absurdly, quite on foot. It is incredible what tricking knaves most stable-people are, and what daring attempts they will make to gain an ascendant over their masters, in order to have their own foolish projects complied with. In shoeing, for example, I have more than once known, that for the sake of establishing their own ridiculous and pernicious system, when their masters have differed from it, they have, on purpose, lamed horses, and imputed the fault to the shoes, after having in vain tried, by every sort of invention and lies, to discredit the use of them. How can the method of such people be commendable, whose arguments, as well as practice, are void of common sense? If your horse's foot be bad and brittle, they advise you to cover it with a very heavy shoe; the consequence of which proceeding is evident: for how should the foot, which before could scarce carry itself, be able afterwards to carry such an additional weight, which is stuck on moreover with a multitude of nails, the holes of which tear and weaken the hoof? If the foot is cut or hurt, one doctor says, load it, by way of cover, with all you can: his conceited opponent as wisely counsels you to let the horse walk bare upon the sore. The only system all these simpletons seem to agree in, is to shoe in general with excessive heavy, and clumsy ill-shaped shoes and very many nails, to the total destruction of the foot. The cramps they annex, tend to destroy the bullet, and the shoes made in the shape of a walnut-shell, prevent the horse's

walking

walking upon the firm basis, which nature has given him for that end, thereby oblige him to stumble and fall, and of course from their shape tear out the nails and ruin the hoof. Feet once got thoroughly out of shape, by the cat walnut-shell, or other ill-shaped shoes, are sometimes irrecoverable, and almost always very difficult to correct; for horn being of a flexible nature, by being confined in a mould, will retain the shape impressed upon it by a bad shaped shoe, which never admits of the natural tread of the foot. The best way, when a horse is thus circumstanced, is to pare his feet down almost to the quick, and short at the toe, and to turn him out without shoes into some soft grass ground 'till the feet grow again before he is shod. They totally pare away also, and lay bare the inside of the animal's foot with their detestible butteris, which must cause narrow heels, because the hard outside of the foot will of course press in, when it finds no resistance, the inside being cut away, and they afterwards put on very long shoes, whereby the foot is hindered from having any pressure at all upon the heels, which pressure otherwise might still perchance, notwithstanding their dreadful cutting, keep the heels properly open, and the foot in good order. The frog

frog should never be cut out; but as it will sometimes become ragged, it must be cleaned every now and then, and the ragged pieces cut off with a knife. In one kind of foot indeed a considerable cutting away must be allowed of, but not of the frog; I mean that very high feet must be cut down to a proper height; because if they were not, the frog, tho' not cut, would still be so far above the ground, as not to have any bearing on it, whereby the great tendon must inevitably be damaged, and consequently the horse would go lame.

The weight of shoes must greatly, wholely indeed, depend on the quality and hardness of the iron. If the iron be very good, it will not bend; and in this case, the shoes cannot possibly be made too light; care however must be taken, that they be of a thickness so as not to bend; for bending would force out the nails, and ruin the hoof. That part of the shoe, which is next the horse's heel, must be narrower than any other, (as is seen in the draught) that stones may be thereby prevented from getting under it, and sticking there; which otherwise would be the case; because the iron, when it advances inwardly

beyond

beyond the bearing of the foot, forms a cavity, wherein stones being lodged would remain, and by pressing against the foot, lame the horse. Broad webbed shoes are very absurd things. Nothing more is wanted, than just iron enough to protect the outward crust of the foot, and to prevent its breaking. The nails in all shoes must, on account of the natural shape of the foot, be driven slanting a little towards the extreme edges of the foot. Any partial pressure towards the inward edge of the shoe, must of course, in a broad webbed shoe, loosen the nails, and consequently tear and damage the foot, supposing even the iron of the shoe good enough not to bend. This inconvenience of tearing out the nails, &c. great as it is, is the best which can happen in this case; for, if the iron was to bend, it would press against the inward part of the foot, and lame the horse just as much as if the shoe had not been bevilled off at all in the proper place, for the picker to be put in, in order to clean out stones, gravel, &c. Making a groove round the edges of shoes, if the iron is not very good, may cause a partial yielding there; but if the iron is good, a groove is very useful, to protect the heads of the nails. Farriers should always examine a

foot

foot before they shoe it, make the shoe, and pierce the holes for the nails further from, or nearer to, the edges of the foot accordingly, as they find the foot requires. The holes for the nails should always be pierced slanting rather outwards. The best way to forge shoes, in respect to the nails, is to make the holes for the nails at twice, with two different instruments: first on the outside of the shoe punch a place, not quite through the shoe, big enough to receive, and cover the head of the nail, when driven in: next punch a smaller hole, from the center of the abovementioned larger one, for the blade of the nail, quite through the shoe: thus the nails are well driven in, protected, and can not be pushed by use too much into the foot, but always keep their firm, proper place, full as well as, nay better than in a grooved shoe in case the iron should not be perfectly good. All shoes should be a little broader at the extremities towards the heels, than elsewhere, except the foot spreads of itself too much at the heel, which is seldom the case; if the horse cuts, they must not be made so: the reason why they should generally be broader there is, that they encourage the foot to grow, spread properly, and therefore prevent narrow heels. It must always

ways be remembered, that where the web grows narrow towards the heel, the feat of the shoe must neverthelefs keep its ufual proper equal breadth within, otherwife the horfe's foot would not have its equal proper bafis, or *appui*, and the shoe would get into the foot, and require frequent removals, which are great inconveniences. The part of the shoe, which the horfe walks upon, should be quite flat, and the infide of it likewife; only juft fpace enough being left next the foot, to put in a picker, (which ought to be ufed every time the horfe comes into the stable, and often on marches) and alfo to prevent the shoe's preffing upon the fole. In fnowy weather, it is particuarly neceffary to pick and clean the feet very often, on marches; otherwife the fnow foon grows very hard in the feet, makes the horfe flip about very much, and hurts him almoft as much as large ftones in the feet would do. Four nails on each fide, hold better than a greater number, and keep the hoof in a far better ftate. The toe of the horfe muft be cut fhort, and nearly fquare, (the angles only juft rounded off) nor muft any nails be driven there; this method prevents much ftumbling, efpecially in defcents, and ferves by throwing nourifhment to the heels, to ftrengthen them;

on them the horse should in some measure walk, and the shoe be made of a proper length accordingly: by this means narrow heels are prevented, and many other good effects produced. Many people drive a nail at the toe, but it is an absurd practice. Leaving room to drive one there causes the foot to be of an improper length, and moreover that part of the hoof is naturally so brittle, that the nail there seldom stays in, but tears out, and damages the hoof. That my directons for shoeing a proper length may be the more clear and intelligible, I have annexed a draught of a foot shod of a proper length, standing on a plain surface, and with it a draught of the right kind of shoe. *(Plate* 16. No. 1. the interior part of the shoe next the foot, and No. 2. the exterior part, which rests on the ground.) Most farriers make shoes thicker at the heels, than at the toes, especially for hard working horses: the great folly of doing so is very easy to be seen, for horse-shoes always wear out sooner at the toe, than any where else; consequently the toe rather requires more substance, than any other part. In some farriers shops the anvils are concave, and the hammers convex, so that it is almost impossible a well shaped flat shoe should be made there. Place the shoe both ways on a flat surface, and it is surprizing how faulty the form of it is generally.

La Fosse's tips,[30] or half shoes, are particularly useful for feet whose crust is too weak to bear nails towards the hinder parts of the foot, and whose heels have a tendency to grow narrow. Pity it is that they require being frequently removed.

In wet, spungy, and soft ground, where the foot sinks in, the pressure upon the heels is of course greater, than on hard ground; and so indeed it should be upon all accounts. The hinder feet must be treated in the same manner as the fore ones, and the shoes the same: except in hilly and slippery countries, where they may not improperly be turned up a little behind: but turning up the fore-shoes is very seldom, I am convinced, of any service, and is very prejudicial to the fore legs, especially to the bullets. In very greasy, wet, or loose kind of slippery soils indeed, where the ground easily gives way, and lets the foot in, without however holding it in very strongly, turning up before may be useful, but in a hard country, where the foot can not enter the ground, cramps before are very hurtful, and quite useless; the tendon being by them elevated, and therefore constantly straining itself for want of a basis to rest on, they endamage the finews very much, and caufe windgalls, lamenefs, fwellings on the bullet, and weakneffes, &c. almoft as much as the walnut-fhell fhaped fhoe, which is held in fuch high efteem by bad farriers, and their ignorant ftable followers. In defcending hills, unlefs in the above-mentioned kind of foils, cramps on the fore feet are apt to throw horfes down, by ftopping the fore legs, out of their proper bafis and natural bearing, when the hinder ones are rapidly preffed; which unavoidably muft be the cafe, and confequently cannot but pufh the horfe upon his nofe. With them on a plain furface, a horfe's foot is always thrown forwards on the toe, out of its proper bearing, which is very liable to make the horfe ftumble. The notion of their utility in going up hills is a falfe one. In afcending, the toe is the firft part of the foot, which bears on, and takes hold of the ground, whether the horfe draws, or carries; and confequently the bufinefs is almoft done, before the part, where the cramps are, comes to the ground. Ice nails are preferable to any thing to prevent flipping, as alfo to help horfes up hill, the moft forward ones taking hold of the ground early, confiderably before the heels touch the ground: they muft be fo made,

as to be, when driven in, about a quarter of an inch above the shoe, and also have four sides ending at the top in a point. They are of great service to prevent slipping on all kinds of places, and by means of them a horse is not thrown out of his proper basis. They must be made of very good iron; if they are not, the heads of them will be perpetually breaking off, which will not happen, if the iron is good, and the nails are well made, of the above-mentioned shape and size. Making them with higher heads, would render them liable to break off, and answer no purpose whatever. When, in the not long ago mentioned kinds of grounds, cramps on the fore feet are used, they should be small, and the heads of the nails should stand up in the manner of the ice nails, but not quite so high, above the shoe, by which the foot and the tendons would always have their proper bearing. These nails may be also used without any cramps. By putting a fresh nail every now and then on the shoe, as wanted, all wished for ends are obtained, and no bad effects ensue. I know that I am fighting against a very strong, though very unreasonable prejudice. Let this method be tried only, and
compared

compared fairly on experience with others; and not immediately laid aside, if, in slippery weather, a horse thus shod should now and then slip. In some weather, and on some ground, any horse any how shod, may sometimes chance to fall. There is unluckily no absolute specific against accidental falling in any shoes yet discovered. I have tried all methods, and find the above-mentioned one the nearest to perfection: this sort of shoe and nails, when well made and fixed properly, being the firmest basis, and best hold I ever knew. I do not recommend ice nails at all times: in certain weather, (the greatest part of the year indeed) the ground is in a condition which does not require any. From the race-horse to the cart-horse, the same system of shoeing should be observed: the size, thickness, and weight of them only should differ: the shoe of a race-horse must of course be lighter than that of a saddle-horse; that of a saddle-horse lighter than that of a troop, coach, draught, or bat horse; and these last more so than a cart, waggon, or artillery horse. A saddle-horse's shoe should weigh thirteen ounces and a half; that of a coach, or draught-horse one pound and three ounces: the nails for the former one ounce per dozen; those for the latter one ounce

ounce and three quarters. Much the easiest way, and in general the best, is to use a narrow-webbed shoe, all over of one equal breadth both within and without, with the holes for the nails exactly in the middle: with little or no art, such a shoe is made out of a narrow bar of iron: it must necessarily be always narrow, for there can be no bevel in it, or it would press upon and hurt the inside of the foot: it has one great advantage over all other shoes, that stones cannot lodge in it. At present all shoes in general are too *heavy:* if the iron is good, shoes need not be so thick, as they are now generally made. With exceedingly heavy loads, such as large cannon, in hilly, slippery countries, and in the bad seasons of the year, the thiller horse[31] should be turned up both before and behind, with three cramps on each shoe; one in the middle part of the toe of the shoe; which in going up hill would help the horse much in his first force to draw his weight after him. I mean this only for a thiller horse, and in certain countries, and weather, when the foot can enter the ground, so that the elevation given to the shoe has no inconvenience attending it. The utmost severity ought to be inflicted upon all those who clap shoes on hot: this

unpar-

unpardonable laziness of farriers in making feet thus fit shoes, instead of shoes fitting feet, dries up the hoofs, and utterly destroys them. It has happened, that the sole has been so much heated by a hot shoe, that a horse has been most dangerously lamed, and some have even lost their lives by it. Shoes should be always made and fitted before the holes are pierced. The shoes in England at present, that are contrived with the most sense, are what they call plates for the race-horses at New-Market :[32] I do not say, that they are perfect, but they are nearer the truth, than any others I know; nor are they substantial enough for common use, though sufficiently so for the turf.

It is sometimes easy to cure horses of cutting by shoeing, but far from always: nine times in ten their doing it proceeds from their turning out their toes. Colts generally graze with one foot stretched out, which rests on the inside, by which the inside is worn down; this makes the toe grow outwards, and the colt becomes crooked from the fetlock downwards: the cutting then generally proceeds from the inside being lower than the outside; the

outside

outside therefore must be frequently pared down, and the inside not. If the foot is such as will not bear cutting, the shoe must be made thicker on the inside web, than on the outside one, from the heel to the toe, and every time the horse is shod, the shoe must be turned a little inwards, and the outside of the hoof rasped off, 'till the foot becomes quite straight by degrees. Bar-shoes can never be good, or useful, but just for a very little time, to cover some damaged part of the foot, if the poor horse can not be spared from work, 'till he is cured.

'Tis strange, that there should be so many ridiculous and absurd methods of shoeing, when it is so manifest, that a small share of common-sense, with a moment's reflection upon the structure of a horse's foot, cannot but suggest the proper one. Frequent removals of shoes are detrimental and tear the foot, but sometimes they are very necessary: this is an inconvenience, which half-shoes are liable to, (though excellent in several other respects) for the end of the shoe being very short is apt to work soon into the foot, and consequently must then be moved. Soldiers should always carry two spare shoes with them, on the upper end and outward side of each holster pipe, with some nails. Some should carry a hammer, others a pair of pinchers, others a butteris, and all be taught how to fix on a shoe. The weight of these things properly divided is trifling. The use of them would be soon found on service, particularly with light troops, and on detachments, where farriers cannot be present.

The common practice of stuffing feet with dung is a very bad one, for the dung contains a rotting quality in it: clay and hog's lard, well mixed together, is much better for that purpose. As to hoof ointment, none is better than that made of one pound of neat's foot oil, one pound of turpentine, and ten ounces of bees-wax. Greasing and stopping, though good for most feet, are not so for all; weak spungy crusts and soles are the worse for it: such must be kept dry. Strong feet must be often wetted, greased, and stopped, and the crust kept down low, or they will fall in by the strong pressure of the crust, and cause narrow heels. When horses are hot, the water with which their feet are washed should be lukewarm: if the heels are cracked, those parts should be washed

with

with milk and water, and a little brandy in it; made a little warm. Mr. CLARKE,[33] in his excellent treatise upon shoeing and feet, insists, that oil, greasy stuffings, and ointments agree but with few hoofs, that they stop the natural perspiration, and that frequent washings with water, moisture, and coolness, keep them in a much more perfect state. The experience I have had since I saw his book, convinces me that he is right in general: the natural and superior benefit which feet and hoofs receive at grass from the dew, rains and moisture of the earth, is a proof of it: and on the other hand we see, that race-horses, particularly at New-Market, where they are always exercised on a dry, close turf, and where they drink out of troughs, round which there is no water for them to stand in, are subject to a variety of diseases in the feet, and hoofs, though they are kept constantly greased.

The methods of treating and keeping horses in other respects, are as various, and for the generality as inconsistent with reason, as those of shoeing are; but a little consideration would (in most common cases at least) direct people right in both. One pampers his cattle, with a view of strengthening them; and afterwards, by way of correction, he pours down drugs into them without thought or measure: another lets no air at all into his stable; from whence his horses inevitably catch cold, when they stir out of it, and are rotted, if they stay in it, by bad corrupted air: a third, equally wise, leaves his stable open, and his cattle exposed to the wind and weather at all times, whether his horses or the weather be hot or cold, and frequently too even in wind-draughts, whilst they are in a sweat. All these different notions and practices are alike attended with destruction to horses; as also are the many extravagances that prevail in the same contradictory extremes, with regard to coverings. But in answer to all these foolish systems, reason plainly suggests to us, that proper wholesome food, a well-tempered circulation of sweet air, moderate and constant exercise, with due care, and suitable cloathing, as weather and occasions may require, will never fail to preserve horses sound and in health.

After working, and at night of course, as also in lamenesses, and sicknesses, 'tis good for horses to stand on litter; it also promotes staleing, &c. At other times it is
a bad

a bad custom; the constant use of it heats and makes the feet tender, and causes swelled legs: moreover it renders the animal delicate. Swelled legs may frequently be reduced to their proper natural size by taking away the litter only, which, in some stables, where ignorant grooms, and farriers govern, would be a great saving of physic and bleeding, besides straw. I have seen by repeated experiments, legs swell, and unswell, by leaving litter, or taking it away, like mercury in a weather-glass.

It is of the greatest consequence for horses to be kept clean, regularly fed, and as regularly exercised: but whoever chuses to ride in the way of ease and pleasure, without any fatigue on horseback, or in short does not like to carry his horse, instead of his horse's carrying him, must not suffer his horse to be exercised by a groom, standing up on his stirrups, holding himself on by means of the reins, and thereby hanging his whole dead weight on the horse's mouth, to the entire destruction of all that is good, safe or pleasant about the animal. No horse's paces can be perfect, nor can he be agreeable, or indeed quite safe, unless his mouth has been made, and his body suppled to

a cera certain degree, so as to be balanced in the rider's hand. A horse's head should be kept high: when it is low, the animal can not be well balanced; for the fore parts being low, and weighing forwards, the hinder parts must of course be high: the fore parts are naturally much more loaded than the hinder ones, though of a less strong construction. The rider ought to know as much as his horse, at least; for, without art, it is impossible to preserve that *union*, and that *together*, if I may so express myself, which are equally pleasing, and necessary: a man on a totally uninstructed horse, or an ill-instructed one, rides, as it were, upon a coach pole.

A great quantity of hay, especially that which is taken from water meadows, or any low and swampy ground, being of a foggy nature, is not good for horses; it hurts their wind very much: it may serve indeed for cart-horses, and for such troop-horses (few of such, thank God, now remain) who are meant for no other use, but to roll on slowly with a fat fellow, full of beer, upon them; who, to the shame of the service, with the badge of soldiership on his back, is a more stupid and lazy animal, than what

he

he is mounted upon, which to its misfortune is rendered so by the sluggishness of its rider. But troops, who are really destined for service, and to be useful, must be active and in wind; the very training them only, to what is absolutely necessary, requires that they should be so, more, or less, according to the different intents and purposes they may be designed for.

Upon service, the allowance of all kinds of forage, whenever there is a possibility of supplying it, is sufficient; but sometimes it cannot be procured for a long while together; besides which misfortune, it is very often most shamefully and carelessly wasted; not to mention, that commissaries in general seldom furnish out the due quantity or quality of any thing, which they have agreed and engaged for, and are most amply paid for.

At home, our horses are crammed and ruined with overmuch hay, and the allowance of corn is scanty. A kind of mill, not to grind corn, but only just to crack and bruise it a little, is so useful, that no regiment should ever march without one. Every grain of it goes to nourishment; none is to be found in the dung; and three feeds of it go further

than

than four as commonly given, which have not been in the mill. Cut wheaten straw, and a little hay too sometimes mixed with it, is excellent food: to a quarter of corn put the same quantity of cut straw, and now and then if a horse is very lean, but not otherwise, about half a one of hay, and let them all be well mingled together; and as chopped straw is generally exceedingly dry, sprinkle a little water upon the feed in the manger. This proportion of chopped straw may seem great, but considering the lightness of it, it is not such in reality. It obliges horses to chew their meat, and is many other ways of use. The quantity of horses food must be proportioned to their size, work, make, appetite, &c.; yet, in regiments it is necessary to fix, and follow some kind of general rule in respect to it. Four of these feeds as above-mentioned, with ten or twelve pounds of hay per day, will be sufficient for most horses on almost on all occasions, except at the piquet late in the year in bad weather; then they should be almost always feeding on something, or other; and if they have no corn, they will consume near forty pounds a day of hay, allowing for some waste, which is unavoidable, especially on bad ground, and in windy weather.

weather. When the forage confifts of unthrafhed ftraw only, eight-and-twenty, or thirty pounds of it for each horfe will do very well, efpecially, if the cutting-box[35] is made ufe of, as it always fhould be. Whenever forage is fcarce, the beft method is to have every thing cut, and given to the horfes every two hours, in nofe-bags, or deep canvafs troughs, fo that the wind may blow none away. Even in time of peace at home, the cutting-box fhould be ufed conftantly. The allowance at home cannot afford fo much, neither indeed is fo much neceffary, when troops are not on fervice. The exercife horfes take at home, though it fhould perhaps be greater, and more conftant, than it is in fome corps, does not require it. A matter of the greateft confequence, though few attend to it, is to feed horfes according to their work, and never to fuffer them to pafs the day quite ftill, without fome motion at leaft. When the work is hard, food fhould be in plenty; when it is otherwife, the food fhould be diminifhed immediately; the hay particularly. Horfes fhould be turned loofe fomewhere, or walked about every day, when they do not work, particularly after hard exercife. Swelled legs, phyfic, &c. will be faved by thefe means, and many diftempers avoided.

I cannot mention the word piquet, without saying something on our pernicious custom of cutting horse's tails entirely off, the inconvenience of which is very glaring in many instances; but in none more, or more seriously so, than at piquets on service, when in hot weather, and in ground where there are many flies. I have often seen our horses, with meat before them, fretting, sweating, kicking about, laming one another, and so plagued with flies for want of tails to brush them away, that they did not eat at all, and so grew out of condition, whilst the neighbouring foreign regiments on the same ground brush'd off the flies with their tails, were cool, quiet, and fed at their ease, and improved. Since that time indeed our cavalry has been ordered to recruit only long tails, and tis to be hoped the nation will follow the example, though old customs, even the worst, I know, are hard to be got the better of. That of cutting off horses tails, ears, and other extremities, is a very old noted one indeed amongst us in England; for so long ago as the year 747, a canon was, by order of Pope Gregory the second,[36] in a letter to St. Augustine, expresly made at an ecclesiastical court in Yorkshire, to abolish, amongst other cruel customs, so barbarous

barous a practice. On duty and marches long tails are very easily tied up properly, and look very well: a nag-tail indeed, suffered to grow a little, protects a horse pretty well. All sorts of grains are foggy feeding, and though they plump up the body, they do not give a wholesome and sound fat: bran too, is not solid food, and is only now and then to be allowed, when horses are heated, to refresh, and open them, if the case requires it.

Whenever hay is put and left in the racks, it should be well cleaned and freed from dust, and not given in too large quantities: in this respect 'tis, like water, much more beneficial, when supplied in small quantities at a time. When a good deal is given at a time, horses spoil, and do not eat the greatest part of it very often, by having blown upon it a good while. A proper quantity of it should be given at twice; a little in the morning before watering, and the rest sometime after they have done their work in the evening. Nothing but good clean wheaten-straw should be left at night in the racks, when the stables are shut up, and the horses left to rest. If hay is left for them, they will frequently stand up to feed almost all night, lie down but little, and take scarcely any rest. A little straw sometimes in the racks during the day time is also proper.

Both before, and after working, horses should be turned about with their croupes to the manger for about an hour. 'Tis a common, but a great error, and very detrimental to horses, to gallop them immediately after drinking; what stable-men call warming the water in their bellies: they ought to be moved only gently. Upon the whole, a very lean horse, and a very fat horse are both in a manner useless to a certain degree: a rough coat is no good symptom; but the means of making it fine should not be by dint of heat and covering, but by dressing and due care. It is of the greatest consequence to a horse's health, that he should always be well rubbed down, and cleaned. Laziness is the true reason why grooms cover horses so much, and keep stables so hot, though they disguise it under the pretence of thinking it wholesome, which indeed however the most ignorant of them really do. A horse when absolutely ruined by over heat will nevertheless very often have a very fine good looking coat.

# BREAKING HORSES, &c.

It is a duty very requisite, and incumbent upon officers, to be as constant, exact, and frequent in going up and down the lines in camp, as through the stables in quarters; and it is likewise adviseable for every one to visit often his own stables, to inspect and superintend the management of the horses. No trimming with cizars should be permitted, but whatever rough hairs appear, should be taken off by dressing. The inside particularly of the ears should never be trimmed, but always kept cleaned: nature has placed hairs within them for reasons very obvious: when they are cut away, dust and insects frequently get into the ears, incomode horses very much, and sometimes cause a serious ailment in those parts. As great inconveniences often happen from horses getting loose, I have affixed a draught and description of the most effectual halter I know of; *(Pla. 17.)* and indeed the only one I have found upon trial, that is capable of preventing it.

This halter has no throat-band, or rather it has, in a manner, two, which are fixed, and begin at No. 1. They cross at 2, are fixed again and end at 3. The nose band is also sowed on at 3. The place 2, where the throat-

bands

bands meet, is a flat button, which is placed, when the halter is well put on, just under the ganaches, (the channel between the two jaw-bones.) The chains, ropes, or leathers, No. 4, which tie the horse in the stable, are also fixed at 3. No. 5, a single cord or leather; if the horse is only fastened with one, which will be as effectual as two.

As horses are generally more supple to the left, than to the right, owing to their being, from their earliest youth, more handled on that side, than the other, they should not only be led with the left hand, in order that they may bend rather to the right, than to the left; but all collars, cavessons, girts, bridles, bridoons, pillar cords, &c. should be made for the same reason, to buckle, and unbuckle on the right side. Horses often hang themselves in their halters, and frequently hurt themselves a good deal by it: the best remedy for such accidents is merely to keep the hurt clean by washing it with lukewarm water with some brandy in it, and every now and then to supple the part with a little green ointment, such as mallows, &c. boiled to a certain consistency, and mixed with sweet oil.

# BREAKING HORSES, &c.

When horses are out of case, have buttons broke out about them, their legs swell, and their coats stare, and there is not time (nor perhaps an absolute necessity for it) to physic them, a rowel,[37] and two ounces of the following powder, given every morning for twenty, or thirty days, in wetted corn, so that none can be blown away, are of great service: the powder to be composed of one pound of liver of antimony, half a pound of sulphur, and a quarter of a pound of nitre, mixed well together: if the horse has a cough, make it into balls, with flour and treacle, or any such kind of thing.

A common complaint amongst troop-horses is broken-wind, which is chiefly occasioned by stuffing them with too much hay; and often by hurrying them too violently after drinking, and after their coming at first from grass. There is no sovereign remedy for broken-wind; but the greatest palliative I know of, is this following receipt of lime-water, which is oftener of service if continued long, or rather always indeed than any other remedy I know of, owing probably not only to the good effects of the lime, but also to the small quantity of liquid the horses take;

for

for very few will ever drink plentifully of this water, and many will go several days without drinking at all, before they will even taste it: the horse must eat no hay at all, and only have wheaten straw in the rack: this water must be used too when mashes are given, and on every other occasion: in short no other water is ever to be given in any shape whatever: 'tis made thus---Take two pounds of quick lime, and put to it twelve gallons of water; mix it over night, stirring it for a long time together, and pouring the water on very gradually 'till the ebullition is over;[38] then leave it to settle for use the next day. If a chalybeate[39] spring is at hand, the lime-water will be much the better for being made of it, instead of any common water. This medicine causes no inconvenience, or impediment, and does not prevent the horse from working as usual. A horse, whose wind is suspicious, should immediately be put on lime-water, and never drink more than a gallon or five quarts in a day, and no horse should drink more than double that quantity, that too at two or three different times. Three pints of warm milk from the cow, night and morning, will sometimes prevent horses heaving, or coughing for a short time, even in tolerably smart exercise; but as

the

the advantages arising from the milk are of so short a duration, this method may, with reason, be looked upon more as a dealer's trick to sell off a broken-winded horse by, than as a remedy. Farriers generally send horses touched in the wind to grass, which, opening them, at first seems to do them good, but, when they are taken into the stable again, and put for some time on hard-meat, they are always worse than before, and the distemper more rooted in.

Worms are so common, and so troublesome a distemper, that I can not omit saying something of them here. Horses, who look out of order, are frequently so owing to worms; that must be examined into always immediately. Give fasting, and let the horse fast three or four hours after it, a quart of beef brine every morning for three or four days. The brine alone will often cure entirely, a purge being given the day after all the brine is taken; a clyster should be given over night, before the purge. If from one ounce and a half to two ounces of Æthiop's mineral in a bolus[40] is given the day after all the brine is taken, and a day before the purge, the cure will be still more certain. You'll see the dead worms in the horse's dung.

A running at the nose, with a cough, and other symptoms, known by the name of *the distemper*, is so frequent, and so ill treated by farriers, that I can not help giving some directions for the treatment of it. Give frequent clysters, keep a rowel or two running for some time, and, if the illness be violent, and attended by a fever, give James's fever powders[41] for three nights running, the first night three papers, the second night two papers, and the third night one paper. No bleeding at first. Then give, for four days running, two ounces of nitre, and afterwards an ounce and a half a day for some time. Poultice from the very beginning under and about the throat, with bread, milk, and lard, made pretty hot; if any thing hard thereabouts grows soft, and does not break of itself, open it with a lancet, and cleanse it thoroughly. As soon as the running at the nose ceases, and not before, give very gentle exercise, and, if the cough then still remains, bleed very little at a time, but frequently, 'till it ceases. Keep the horse by no means cold, but let him have fresh air. He must not be moved 'till the running at the nose ceases. Don't physic, but continue the ounce and a half of nitre for three weeks at least, and give two or three times a week,

for

# BREAKING HORSES, &c.

for as long as is found necessary, a drink made of liquorice root, stones of raisins bruis'd, and figs dry'd, of each two ounces, and one ounce of maiden-hair; boil them together in a quart of water, 'till reduced to a pint, then add syrup of balsam, cold drawn linseed oil, of each two ounces, and one ounce of nitre. This drink not to be given 'till the running at the nose ceases. If the distemper is exceedingly slight, James's powders, may be omitted. If the testicles swell, use cooling things, such as warm milk and water, marsh-mallows, &c. but above all things, don't neglect to suspend them in a sling. Keep the nose and nostrils very clean, by washing them frequently with warm water. Feed with mashes only, and continue the poultice 'till the running of the nose has ceased two or three days. Then the covering about the throat must be taken off by degrees, a little at a time.

Greasy and swelled legs being a very common distemper in troop horses, I shall set down the following very good receipt for the cure of it :---Take salt-petre two ounces and two drams, the same quantity of venice turpentine, one ounce and four drams of flour of brimstone, diapente six drams; mix the whole together with a sufficient quantity of liquorice powder, make it into balls, and give it to the horse fasting in the morning; he must not eat for two hours after taking it, nor drink for five or six hours, and then the water must be warmish; he must be kept warm, and have gentle walking exercise the next day; this dose must be repeated twice, or more, as required, with an interval of three days between each dose.

The following manner of treating the grease is also a very good one.---As medicines to be given inwardly, take of powdered resin one ounce and a half; of salt of tartar, and sal prunell, each six drams; spirit of turpentine, enough to make it into a ball. The proper dose for a large horse is three ounces: it should be given when first made up, or else the salt of tartar will make its escape. This will operate as a diuretic two days, during which time the horse is to have plenty of scalded bran, plenty of warm water, and gentle walking. The third and fourth morning, he is to take a ball made of the following medicines. Take of foenugreek, aniseed, elecampane, turmerick, liquorice powder, diapente powdered, each equal parts; add to a pound of this powder two ounces of anisated balsam of

sulphur,

sulphur, and honey enough to make it of a proper consistence: the dose of this ball to be of the size of a hen's egg: the diuretic ball is to be given in the morning; the day following nothing; the two succeeding mornings, the cordial ball; and so on 'till the diuretic ball has been given three times: the cordial ball to be continued every day after the third diuretic ball is given, 'till the horse is well.

As external applications,---if there be a swelling of the parts, they should be poulticed with warm rye meal, and milk, boiled to a proper consistence, which is to be renewed every day. When the swelling is gone, apply the following: take of honey two pounds and a half; of train oil, and powdered allum, each two pounds; boil them to a proper consistence: some of this to be spread on a linen rag, and applied to the parts: to be renewed once in forty-eight hours. The horse must not go out, when this medicine is applied. This will dry up the sores, and, if there is any scurf, or scab left, use the following mixture: take of the juice of houseleek one part; of very thick cream two parts; beat it up together into an ointment, and rub some of it every day on the parts affected.

Resin

Resin drink is also very good for swelled legs. The following is also a good method of curing the grease: pluck out the hairs clean, with pinchers, all about, and upon the greased part. Then put on a turnip poultice, and leave it on twenty-four hours; then spread a linen bandage with tar, and wrap it, not loose, nor tight, round the part, and leave it on three or four days. Continue at the same time, the balls, or resin drink, and take away some blood once or twice, a little at a time.

When a horse is lame, no matter where, grooms and farriers generally say he is so in the shoulder, which is very seldom the case. If he really is so, he will drag his toe on the ground, or move his legs circularly, more or less, according to the degree of the hurt; if he does not do it at all, he is not lame in the shoulder. Every body who is in the least acquainted with the texture of a horse, knows this to be true. When a horse's lameness proceeds from any other cause, from the knee downwards, one may generally know it by some inflammation, or other sign, such as swellings, tendernesses, &c. One may generally suspect with reason something wrong in the feet, or coronary ring, owing chiefly to the common very bad method of managing feet. Running thrushes are

are a common complaint, and though they are to be stopped, generally end in eating away the inside of the foot: Vitriol and water dry these thrushes, and so does a mixture of one-third spirit of nitre, and two-thirds of spirit of wine dabbed with a rag, and several other applications of that kind. When horses, who are troubled with them, tread on a sharpish stone, the pain they feel from it is often so great, that they fall down as if they were shot. Sometimes a clumsy fellow, by negligence and aukwardness, which is oftener the case, than by any other accident, is the cause of his horse's falling, and breaking his knees. If any thing will make the hair come again, and probably of a right colour, burnt cork finely sifted, mixed with oil, and made into an ointment will do it; but if the horse is grey, the burnt cork must be omitted, and honey mixed up with the oil in lieu of it, because the burnt cork, by causing the hair to grow up of a darkish colour, would disfigure a grey, or white horse. Before the cork, and oil ointment is used, poultice the part with pounded turnips boil'd with milk, and mixed up with hog's lard, and a little friar's balsam; 'till there is no swelling or irritation left. The poultice must be put on fresh every twenty-four hours; the ointment must be laid on very often, and the part must be kept free from dirt.

For

For strains of all kinds, soap, and camphor dissolved into spirits of wine, and often well rubbed on the part, which must be afterwards covered with tow and warm pitch, are excellent. The tow thus stuck, and left on, keeps the injured part from cold, &c. and it is some time before it wears off: it is indeed a blemish for the time, but besides being a good remedy in itself, it is otherwise of great use, as it puts it absolutely out of the power of grooms and farriers to play any of their tricks, or for the latter to have any pretence whatsoever to be about the stables. It is a common custom to give walking exercise to horses who have sprains, which is very pernicious; they should not be stirred at all, if possible: absolute rest is the best remedy for them.

A blanket for each man carried under the saddle is of vast use to the horse's back, as well as to the man on many occasions. Every man should have one.

Every troop ought to have a cutting-box belonging to it, and one man constantly employed in camp all day at it in chopping hay, straw, &c. It is very easily carried about.

Forage, whatever it is, must not be cut too long, nor very short,

short, but of such a length, that it may not, from its lightness, be blown up the horse's nostrils out of the nose-bag, or canvass trough. A lazy fellow at the cutting-box, if not watched, is very apt, by way of getting rid of his work soon, to cut it much too long.

The Germans wisely carry, upon all occasions whatever, every man a double feed of chopped straw and corn mixed together, which is never touched, but by express order of the commanding officer, and then too in such quantities, and at what time, he thinks fit to direct. It frequently happens upon long marches, and even sometimes when the troops stand still, that forage cannot be procured for some days together; then this practice, which I have just mentioned, in a short time gives strong and apparent proofs of its utility, by the preservation of their horse's good plight. It is the means of saving the lives of many horses, and helps, in cases of exigencies, to keep up the vigour of most of them. None but those, who have been eye-witnesses to the fact, can tell what harm a deficiency of forage, only for two days, does horses, especially in marches by night, and in bad weather: some are often disabled by it for the whole campaign, and some for ever after.

In the beginning of September, in our climates, green forage is no longer plenty on the ground. It would therefore be prudent from that time to make every man carry twenty pounds of spun hay, and afterwards later in the year a larger quantity. From about the twentieth of September, for example, or thereabouts, he might carry thirty pounds for the rest of the campaign, and, besides this hay, eight pounds of oats mixed with four pounds of cut wheaten straw, none of these to be ever touched, but by order of the commanding officer, and then in such quantity as he thinks fit. This method would often prevent troops from being in great want, and richly repay the horse for carrying the forage. As hay spoils by being kept twisted up for a long time together, it should be unspun, and given to the horses at the end of three days, and a fresh truss spun, and made up. If the campaign should last through the whole winter, this forage must be carried, 'till there is green forage enough on the ground the ensuing year, which may not be 'till late, in poor uncultivated countries, or those worn out by war. Whenever horses come out of quarters, where they have met with abundance, corn must be taken from them by degrees, if possible, and not all at once, be the season, and the country they take the field

in

in ever so good. For a considerable time horses will do very well in the field without corn, if, on coming out of quarters, they are not weaned from it too suddenly, and the weather, and green forage is tolerably good; but late in the year, when the weather grows bad, and horses are obliged to go a great way for forage, some corn is absolutely necessary.

In fetching forage, especially from any distance, the trusses should be very well made and fixed, and no men suffered to ride on them; the weight of both being immense. I have very often seen trusses of three hundred weight, which without a man on it, is a very heavy load. Laziness and custom has made some people imagine that a truss of forage cannot be carried without a man on it, but it is not so by any means, if the trusses are well made, and properly fixed. These, and many other precautions and care, in matters, seemingly perhaps little and trifling, ought to be deemed, (as they really are) equally as necessary for preserving a regiment in the condition it ought to be for its own credit, and the public service, as a just distribution of rewards and punishments. These, and such-like attentions should no more be dispensed with, than that an officer of each

troop

troop should constantly visit every horse of that troop daily in their lines, cantonments, or quarters; and especially too, and without delay, after fatiguing marches, and foul weather: but if this care be intrusted to a quarter-master, who is already over-loaded, not only with his own, but often with the whole business of the officers, beyond a possibility of executing half of it; and if he likewise, (being indeed in some measure compelled to it) shuffle off his burden, all he can, upon the serjeants and corporals, what else can be expected, but that the same spirit of idleness and disregard will diffuse itself throughout the whole corps? Hence no duty would be compleatly and essentially performed; none in the stables or camp with respect to the horses, accoutrements, &c. no regularity in cooking; no care to see the men well dried after wet service; in short, no serious attention to numberless other necessary articles of discipline, &c. whereby a regiment would most infamously fall to ruin, and be very soon rendered unfit for service.

**THE END.**

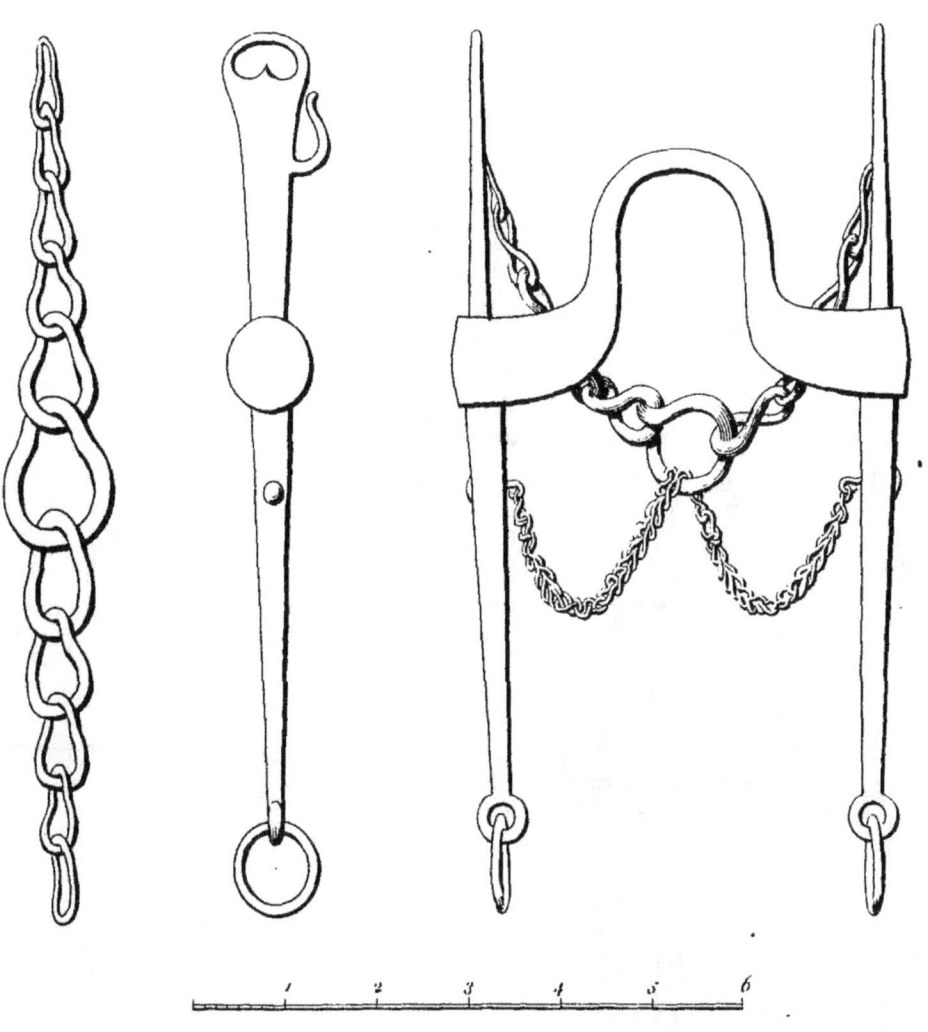

*Pla.1.*

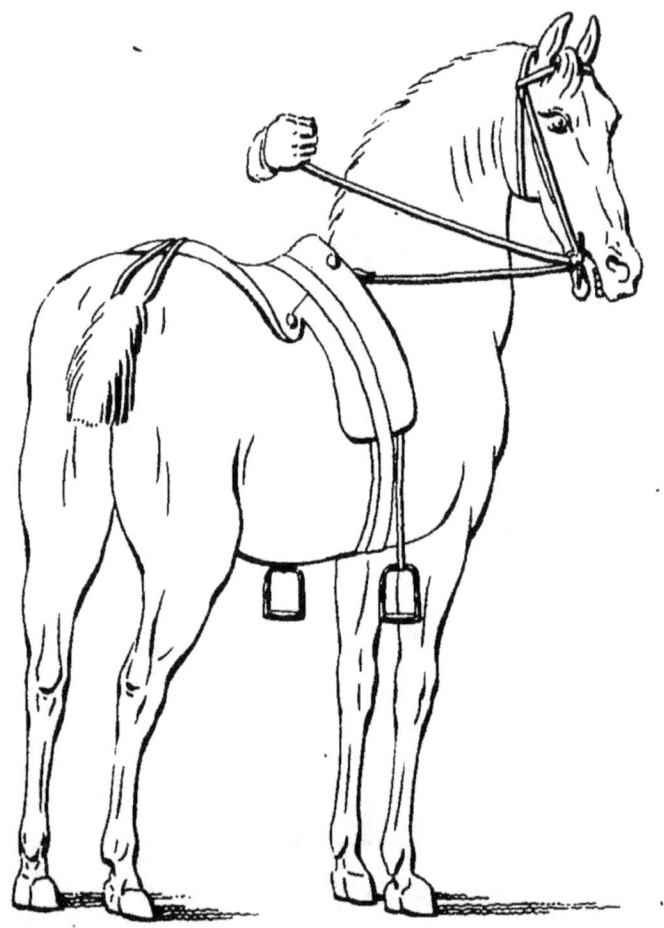

Pla. 2.

Pla.3.

Pla. 4.

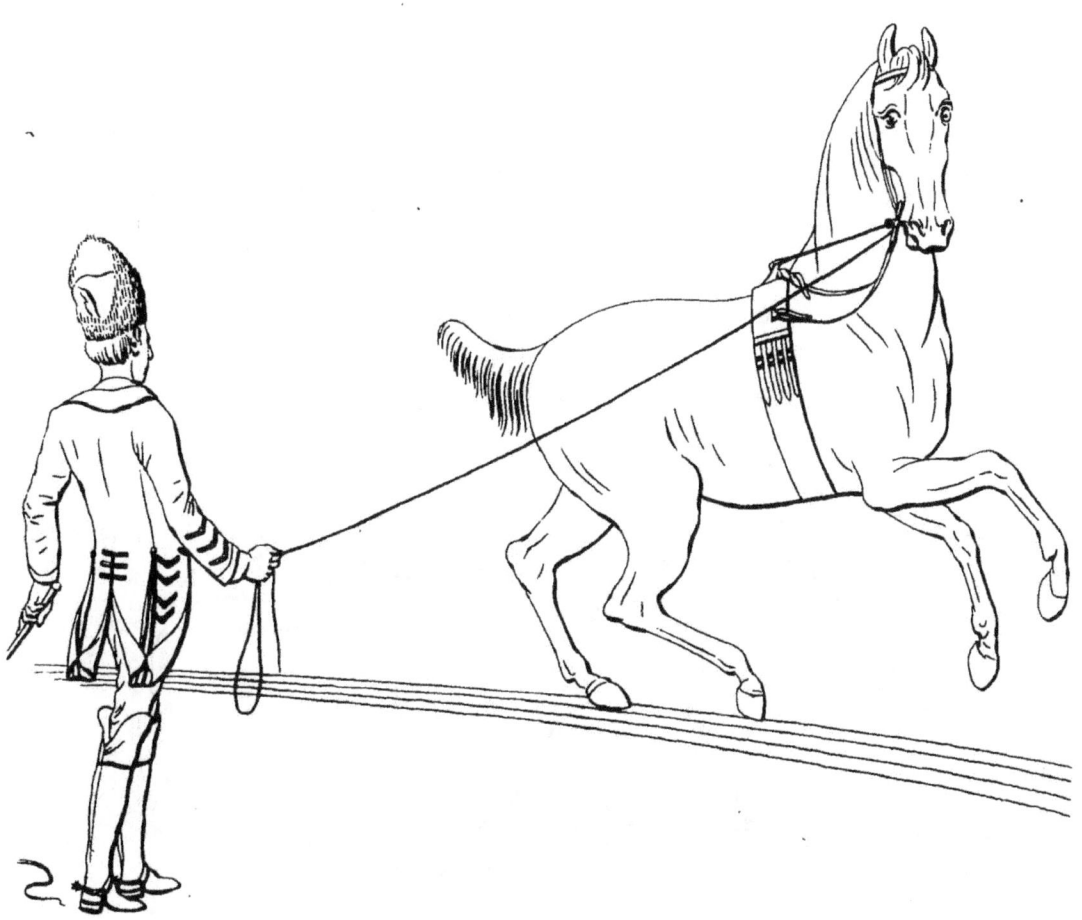

Pla.5.

Pla. 6.

Pla.7.

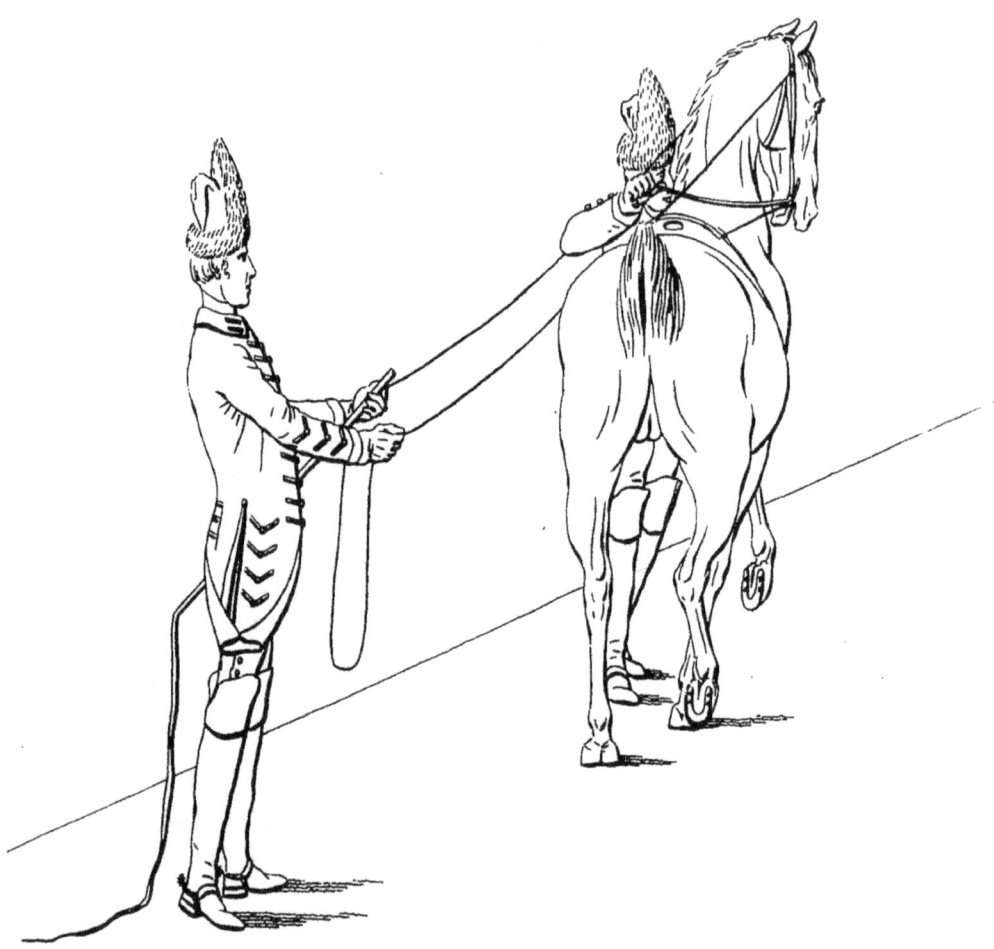

Pla. 8.

Pla. 9.

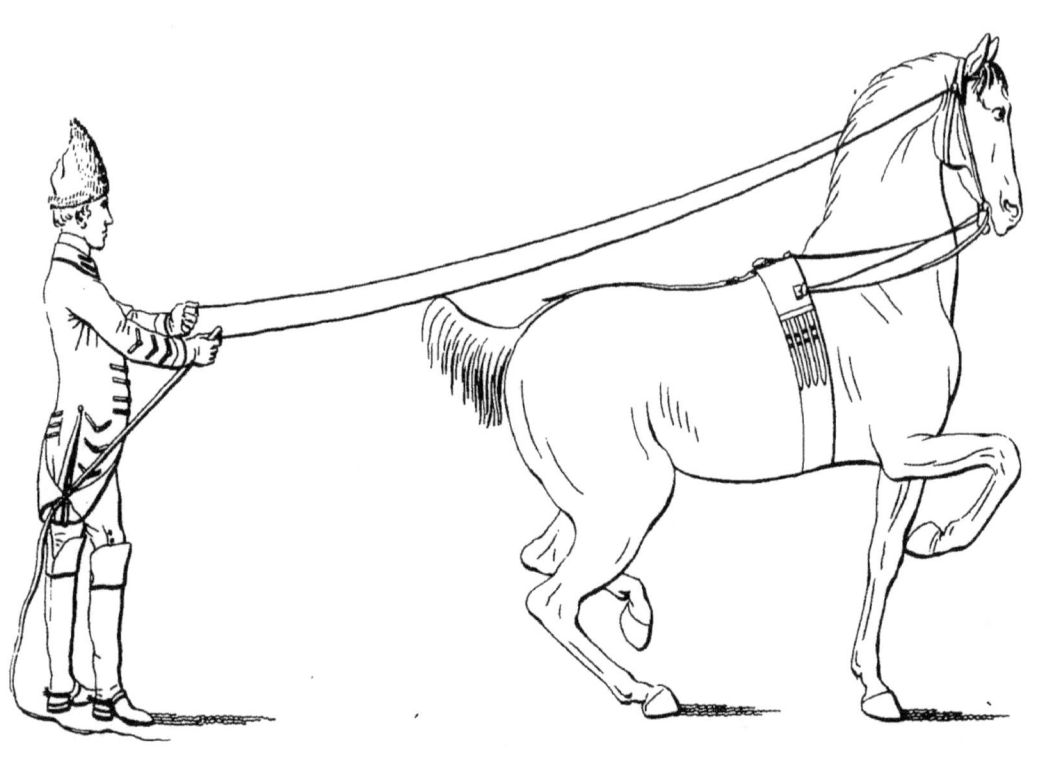

Pla. 10.

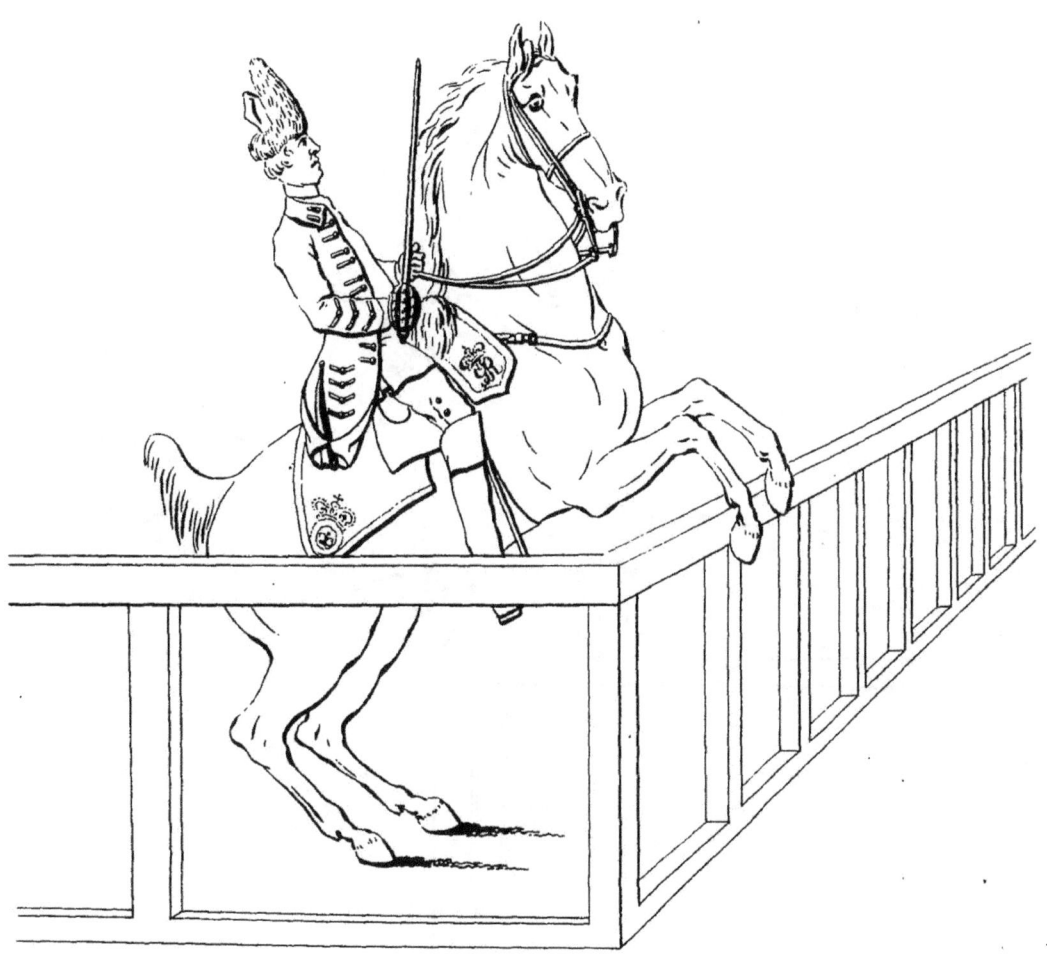

Pla. II.

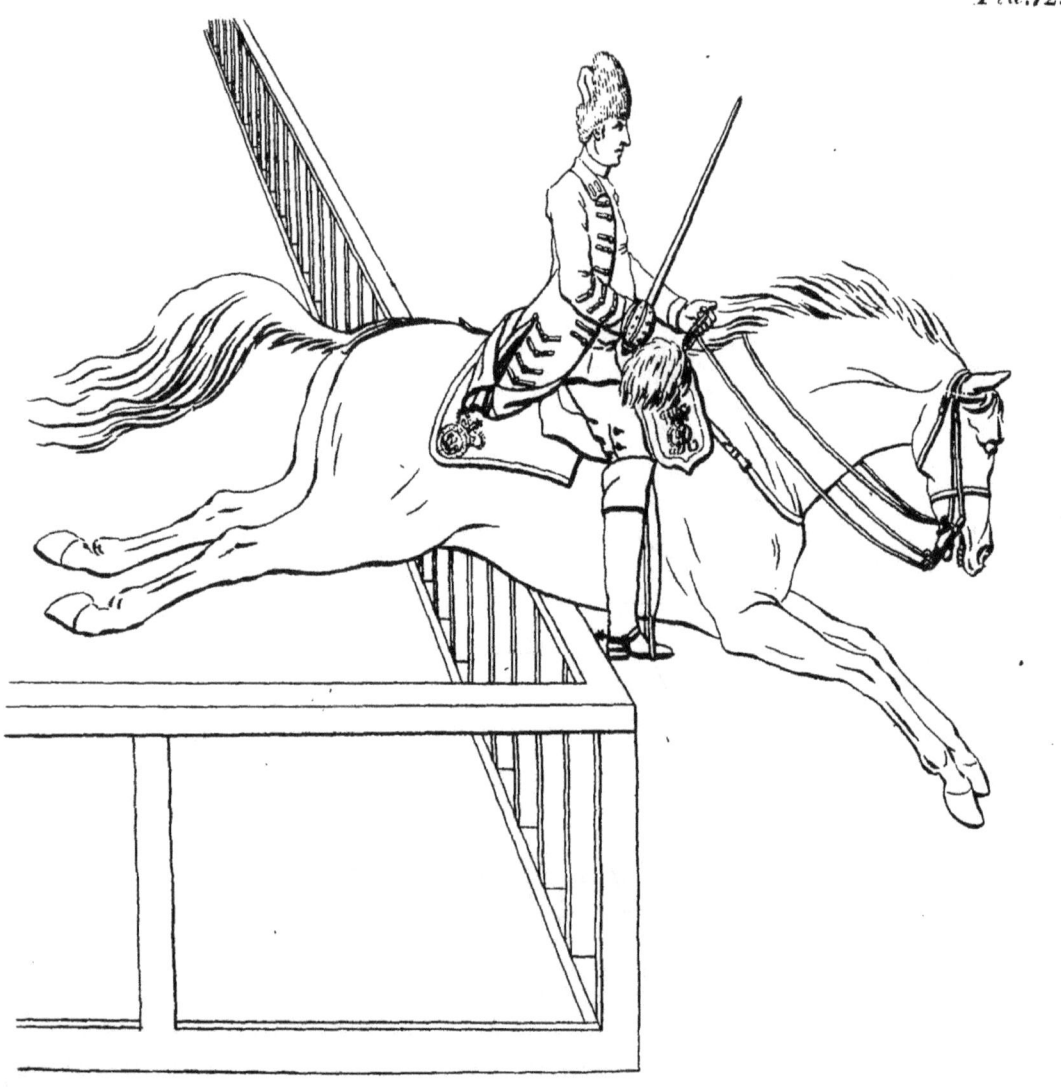

Pla. 12.

Pla.13.

Pla. 14.

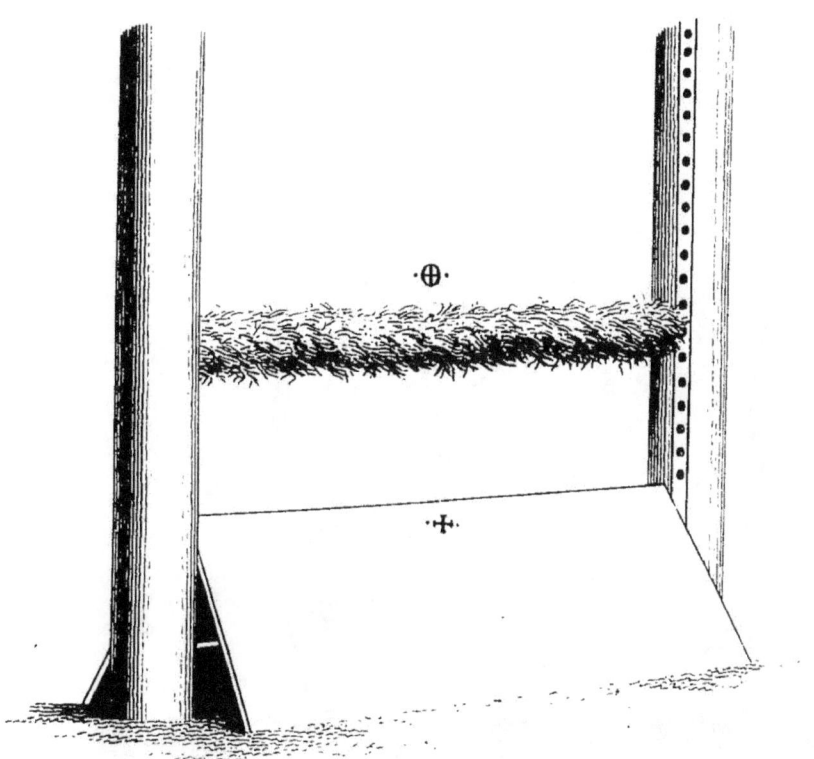

Pla. 15.

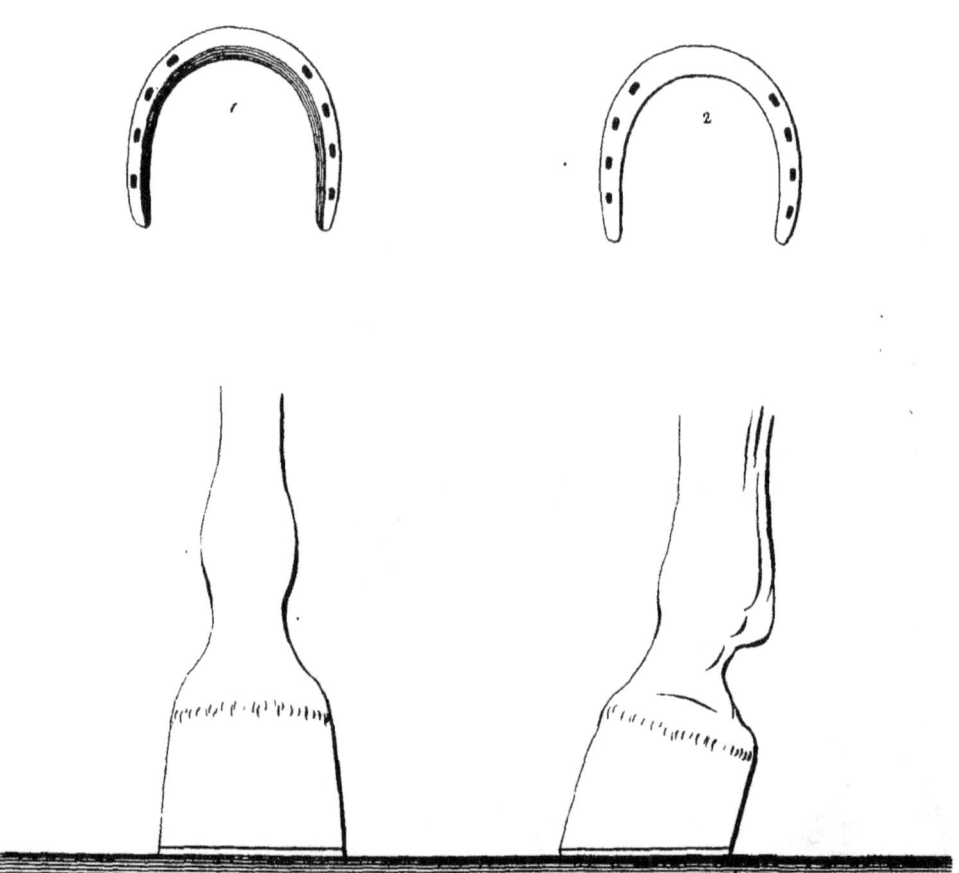

Pla. 16.

Pla.17.

# EXPLANATORY NOTES TO A METHOD OF BREAKING HORSES, AND TEACHING SOLDIERS TO RIDE

## Title page

[1] *Scientia, & Patientia.*

Knowledge & Patience.

[2] _____  _____*Equitem docuere sub armis*

*Insultare solo, et gressus glomerare superbos.*     Virg.

And taught the rider under arms to paw the ground, and curvet in proud ambling pace. (*Virgil, Georgicon*, 15)

[3] *Vis consilî expers mole ruit suâ.*                              Hor.

Rash Force by its own weight must fall. (*Horace, Odes, Satyrs, and Epistles*, 84)

## Dedication

[4] General George Augustus Eliott. George Augustus Eliott, first Baron Heathfield of Gibraltar (1717-1790), "led a regiment of light cavalry throughout the campaigns in Germany during the Seven Years' War between 1759 and 1761 [and] took a prominent part in the bold cavalry charge at the battle of Emsdorf (16 July 1760)." Following the war in 1763, Eliott's regiment was titled the King's Own Royal light dragoons, subsequently the 15th hussars. Eliott was created Baron Heathfield of Gibraltar in 1787 (Falkner, ODNB).

**Page 17**

⁵ *flat, or demi-piqued saddles.* Piqued saddles with high pommels and cantles that essentially encase an armored rider were used by heavy cavalry. Semi- (or demi-) piqued saddles with flatter profiles provided more freedom of movement and were used by light cavalry. As a later treatise from 1893 notes: "Before hunting habits put high-school precepts out of fashion our ancestors rode either upon the high piqued saddles used by knights in armour, or on a demi-pique something like the modern military" (Sidney, 304).

**Page 21**

⁶ *Appui* can be defined most simply as "contact." Grisone introduced the concept *(appoggio)*, and Pluvinel, Cavendish, and de la Guérinière all discussed it *(appui* or *appuy)* at length. The Dictionary appended to the 1743 English translation of Cavendish's General System of Horsemanship defines *appui*, "or stay upon the hand, [as] the reciprocal sense between the horse's mouth and the bridle-hand, or the sense of the action of the bridle in the horseman's hand"; and de la Guérinière calls it, in apposite terms, "the sensation produced by the bridle on the hand of the rider, and conversely, the action which the rider's hand communicates to the bars of the horse." Horsemen have spent half a millennium debating the theory and practice of this foundational concept in dressage and equitation (Cavendish, np; de la Guérinière, 90).

**Page 30**

⁷ *Carabine* [or Carbine, page 82]. "Carabine, or Carbine, a small kind of fuse or fire-arm, about two feet long in the barrel" (RED).

**Page 41**

⁸ *Epaule en dedans.* Shoulder-in. Described by Pembroke as "a very touchstone in horsemanship," the invention, or at least development, of the shoulder-in is generally attributed to the 18th century master François Robichon de la Guérinière, who devoted a full chapter of *School of Horsemanship* to what he termed "the alpha and omega of all exercises for the horse which are intended to develop complete suppleness and perfect agility in all its parts" (de la Guérinière, 137).

### Page 42

[9] *Sir Sidney Medows.* Sir Sidney Medows (c. 1699-1792), Knight-Marshal of the Marshalsea-court in Southwark, was eulogized as, "perhaps, the most complete rider of managed horses in the kingdom; and so fond of it was he to the last, notwithstanding his great age, that, not many hours before his death, he made his servants set him on horseback" (Urban, 1236). Richard Berenger dedicated his translation of Claude Bourgelat's *Nouveau Newcastle* (see note to page 61 below) to Medows, and, in *The History and Art of Horsemanship*, Berenger places Medows in Newcastle's company but notes: "He never yet has thought proper to convey his knowledge to others by means of the Press, but . . . does more than other people write. His Horse is his Pen . . . (Berenger, I, 213). Medows's knowledge was conveyed posthumously by his pupil Strickand Freeman in *The Art of Horsemanship Altered and Abbreviated, According to the Principles of the Late Sir Sidney Medows* (1806).

### Page 42

[10] *Cavaliere Rossermini* [sic], *at Pisa, author of the Cavallo Perfetto.* Niccolò Rosselmini (?-1772), was a "patrician of Pisa, chamberlain of the Grand Duke of Tuscany and superintendent of the stud-farms of the Grand Duke in San Rossore, as well as director of the riding school of Siena" (Tomassini, 236). His equestrian works include *Il cavallo perfetto.* Venezia: Giuseppe Corona, 1723; *Apologia del cavallo perfetto*, Siena: Francesco Quinza, 1730; and *Dell'obbedienza del cavallo.* Lavorno, Marco Coltellini, 1764 (Tomassini, 236-40).

### Page 44

[11] *Chambriere.* Long whip, as translated from de la Guérinière by Tracy Boucher (de la Guérinière, 124). The word appears only in its sense of chamber-maid or hand-maid in James Howell's *French and English Dictionary* (1673) and A. Boyer's *Royal Dictionary, French and English, and English and French* (1729).

### Page 61

[12] *Monsieur Bourgelat's Nouveau Newcastle [and] Mr. Berenger's translation*

Claude Bourgelat (1712-1779) was a pioneer of veterinary medicine, founder of the first veterinary school (in 1762 in Lyons, France), author of seminal medical works such as *Art Vétérinaire, ou Médecine des Animaux* (1767), and an avid and accomplished horseman. His *Nouveau Newcastle, ou Traité de Cavalerie* (1747), was a distillation of the Duke of Newcastle's first manual on horsemanship, *La méthode nouvelle et Invention extraordinaire de dresser les chevaux* (1658), whose second edition recently had been published in 1737 and translated into English in 1743 (Chalmers, 230-31).

Richard Berenger (bap. 1719, d. 1782), a courtier and equestrian, translated Bourgelat's *Nouveau Newcastle* as *A New System of Horsemanship, From the French of Monsieur Bourgelat* (1754), and wrote a two-volume *History and Art of Horsemanship* (1771), which included his translation of Xenophon (see note to page 88 below). He was appointed Gentleman of the Horse to His Majesty in 1760 (Courtney, ODNB; Van der Horst, 564).

[13] Pembroke is referring to the diagonal walk which is caused by collecting the walk. The concept of a pure, four-beat walk was more widely recognized in the 19th century. The diagonal (very collected walk) led easily to the training of piaffe, and the parade trot. – *Publisher's note.*

## Page 65

[14] *Ramingue.* Howell's *French and English Dictionary* (1673) defined a *cheval ramingue* as "an inconsistent moving horse, one that keeps not in any direct way; one that holds not any certain or settled order in going" (Howell). Pembroke applies ramingue to horses inclined "to retain themselves, and to resist by so doing."

## Page 78

[16] *Dressing* [a horse]. To train, from French *dresser*, hence dressage.

## Page 83

[16] *Light cavalry.* See Introduction, note 6.

[17] *Squadrons.* "A Body of Horse, the Number not fix'd, but from an hundred to two hundred Men, sometimes more, and sometimes less, according as Generals see fit, the Army in its Strength, and Occasion requires" (*A Military Dictionary*).

[18] *Lines.* "Is the drawing up of an Army for Battle, extending its Front as far as the Ground will allow, that it may not be flank'd. . . . Armies generally draw up in three Lines; the first call'd the Van; the second the Main Body; and the third the Reserve; with a convenient Distance between them, and Intervals, that they may not put one another into Confusion" (*A Military Dictionary*). In *Instructions for Young Dragoon Officers* (1796), William Tyndale speaks of lines and columns as "different shapes an army can assume . . . It most generally assumes [the line] when acting on the offensive or defensive, . . . [and] the column . . . when in motion" (Tyndale, Instructions, 21-22).

## Page 84

[19] *Furze.* A plant which grows wild on heaths and uplands commons, generally used for fuel, or making hedges (RED).

## Page 86

[20] *Partizans.* Partizan or partisan was used as early as 1692 to mean "a member of a small body of light or irregular troops operating independently and engaging in surprising attacks" (OED). In Pembroke's example, partizans are a detachment of light cavalry on "reconnoitring duty."

[21] *Numidians.* "Numidia, under the Roman Republic and Empire, a part of Africa north of the Sahara, the boundaries of which at times corresponded roughly to those of modern western Tunisia and eastern Algeria. Its earliest inhabitants were divided into tribes and clans. They were physically indistinguishable from the other indigenous inhabitants of early North Africa and, at the end of the Roman Empire, were often categorized as Berbers. . . . Numidian horsemanship, animal breeding, and cavalry tactics eventually contributed to later developments in Roman cavalry. In his history of Rome, Polybius underscores how important those cavalry advantages were to the outcome of the Second Punic War" (Editors, *Encyclopaedia Britannica*). In *A History of Cavalry from the Earliest Times* (1877), Colonel George T. Denison writes that "the Numidian cavalry, which formed the light horse of Hannibal's army, were reputed to be the finest light cavalry of the age. There are such conflicting accounts as to their armament and equipment, that it is difficult to describe their method of fighting with much certainty" [followed by such a description] (Denison, 47-48).

### Page 87

[22] *Xenophon.* The Athenian historian and soldier Xenophon (c. 430-354 BC) wrote two seminal books on horsemanship: *On Horsemanship* (also known as *The Art of Horsemanship*), dealing with "the selection, care, and training of horses in general"; and *The Cavalry Commander*, dealing with "military training and the duties of the cavalry commander." Translated into Italian and published in 1580, Xenophon's works were among those in many intellectual disciplines recovered from *Antiquity during the Renaissance* (van der Horst, 200).

### Page 90

[23] *Ceteris paribus.* Other things being equal, other conditions corresponding (OED).

### Page 92

[24] *Duke of Newcastle.* William Cavendish, 1st Duke of Newcastle upon Tyne (1592-1676), an agile and erudite polymath, enjoyed renown as an equestrian and trainer of horses and riders, philosopher and scientist, poet and playwright, music scholar and musician, and diplomat and politician, among other endeavors. Landed, wealthy, and influential at court, Cavendish was a monarchist appointed General in the loyalist army during the English Revolution. Following the execution of Charles I, Cavendish fled to the Continent, and "spent the greater part of his time abroad in training horses," refining the art of manège, and establishing an important riding school at Antwerp. Following the Restoration of the Crown in 1660, Cavendish returned to England, was created Duke of Newcastle, retired from public life, and devoted himself to horse training and breeding (see Hulse).

Newcastle published two manuals on horsemanship: *La méthode nouvelle et Invention extraordinaire de dresser les chevaux* (1658), translated into English in 1743, and A New Method, and Extraordinary Invention, to Dress Horses (1667), which Newcastle described as "neither a translation of the first [manual], nor an absolutely necessary addition to it." After an initial French translation of the latter work in 1671, Jacques de Solleysell published a second translation, approved by Cavendish, in 1677 (adapted from van der Holst, 312).

De la Guérinière, who invokes Cavendish repeatedly in School of Horsemanship, notes that he was "considered to be the greatest expert of his age in the matter of horses" (de la Guérinière, 78).

## Page 95

[25] *argumentent de ces systèmes déplorables.* argue about these appalling systems.

## Page 98

[26] *Forge-cart.* "A travelling forge for service in the field" (OED). A forge-cart carried "shoes, nails, a forge and fuel, an anvil, and the farrier's hand tools, [including] a stand for the hoof to rest on while it was being rasped" (Macdonald, 83).

[27] *made by the Hanoverian train.* This phrase appears in the edition of 1761, with more clarity, as "made by some belonging to the Hanoverian train." Evidently, Pembroke's forge-cart was built by personnel in the Hanoverian artillery train, whose significant establishment in the Seven Years War (1756-63) included roughly 3,000 horses, 1,000 drivers, and 700 vehicles. Britain and Hanover were allied in that conflict (see "Hanoverian Army Train").

## Page 99

[28] *Physic and a butteris* . . . In medicine, a physic was a purge administered to a human or animal. In farriery, a butteris[s] was "an instrument of steel, set in a wooden handle, used in paring the foot, or cutting the hoof of a horse"—Pembroke's call for its banishment was echoed by Peter Bedford in Thoughts upon Hare and Fox Hunting (1781) (OED).

[29] *clyster.* "In Medicine, a decoction of various ingredients injected into the anus by means of a syringe, or pipe and bladder" (RED).

## Page 107

[30] *La Fosse's tips, or half shoes.* Philippe Etienne Lafosse (1739-1820), descended from an old family of farriers, was the fourth and last in a direct line holding the office of maréchal at the royal stables in Versailles. (His father, Etienne-Guillaume Lafosse, in addition to holding that office, wrote "a number of texts on horse diseases," including Traité sur le veritable siege de la morve de chevaux, et les moyens d'y remédier [1749]). "The most famous member of the Lafosse family," and the enemy and rival of his

former protégé Claude Bourgelat, Phillippe Etienne "produced several works on the subject of horses, especially on horse care, medicine and anatomy," including Guide du Maréchal (1766), "his first substantial work on the horse in general, and horse medicine and horse shoeing in particular." His "chief work," the lavishly illustrated anatomical treatise, Cours d'hippiatrique, or traité complet de la medicine de chevaux (1772), "is generally considered as the most magnificent work ever produced in the history of veterinary literature." (van der Horst, 620-23; and see Clerc, 36-37)

"The first modern writer who attempted to reform the common mode of shoeing, appears to have been Lafosse. . . . The shoe recommended by Lafosse was what he called the half-moon shoe, being nearly semicircular, and reaching little further than to the middle of the foot; the nails being placed round the toe. . . . It has been considered as useful in some cases of diseased feet, and for strong feet which have begun to contract, or appear likely to do so, provided such horses are not employed on very hard, rough roads; it is by no means applicable to the majority of our [British] horses" (Kirby, 448). Kirby proceeds to compare and contrast Lafosse, Osmer, Pembroke (at some length), and Clark. Kirby's article on Farriery (a subject first addressed in the Fourth Edition of the *Encyclopedia Britannica* in 1810), runs to 155 pages.

## Page 111

[31] *thiller horse.* "Thill, the shafts, or arms of wood between which a horse is placed in a carriage: hence thill, or thiller horse, the horse that goes between the shafts" (RED).

## Page 112

[32] *race horses at New-Market.* "Newmarket (Heath): Famous English racecourse, first used for hare hunting in the seventeenth century by King James I. Used for 'matches' by King Charles II in 1684. Today known as the Headquarters of the English Turf" (Bloodgood and Santini, 136).

## Page 115

[33] *Mr. Clarke, in his excellent treatise upon shoeing and feet.* James Clark[e] (dates unknown) was Farrier to His Majesty for Scotland, and, like Bougelat, LaFosse, and many early farriers, was both a medical practitioner and horseshoer. Kirby writes that "the shoe

recommended by Mr Clark did not differ very much from that of Osmer. He does not, however recommend the hollowing of the surface of the shoe next the foot . . . and [like Osmer] was much against raising the heels with calkins" (Kirby, 448).

Presumably, Pembroke is referring to Clark's *Observations upon the Shoeing of Horses: Together with A new Inquiry into the Causes of Diseases in the Feet of Horses* (1775). Clark was also the author of *A Treatise on the Prevention of Diseases Incidental to Horses, To Which Are Subjoined Observations on Some of the Surgical and Medical Branches of Farriery* (1788).

## Page 120

[34] *Piquet.* "In fortification, sharp at one end, usually shod with iron, used in laying out ground and measuring its angles; or driven into the ground by the tents to tie the horses to; and likewise used to fasten the cords of tents; whence to plant the picket, implies to encamp" (RED). Hence Picket, or Piquet Guards. "Small Guards commanded by Lieutenants, or Ensigns, at the Head of every Regiment, as they lie encamp'd, to be always in Readiness against all Surprizes" (*A Military Dictionary*; and see Hinde, 491-98).

## Page 121

[35] *Cutting-box.* "A chaff- or straw-cutter." The first usage cited in the OED is from W. Ellis, Mod. Husbandman, January 1744; the second is from Pembroke, page 136.

## Page 122

[36] *Pope Gregory the second, in a letter to St. Augustine.* Saint Gregory II (669-731) was pontiff of the Holy Roman Empire from 715 through 731. St. Augustine doubtless refers to Saint Augustine of Canterbury (?-604/05), dispatched to England by St. Gregory the Great (Saint Gregory I) (540-604) in 596. Consecrated bishop of the English in 1597, Augustine became the first Archbishop of Canterbury (Editors, *Encyclopedia Britannica*). Pembroke appears to be confusing Gregory II and Gregory I and also mistaking a date.

## Page 127

[37] *rowel.* Pembroke is not referring to the rowel of a spur, but to "a thin piece of leather or other material, typically circular with a hole in the centre, inserted into an incision

between the skin and subcutaneous tissue of a horse or other animal in order to produce a discharge," a usage found in Blundeville, Markham, and Pembroke (OED).

## Page 128

[38] *ebullition.* "The process of boiling, or keeping a liquid at the boiling point by the application of heat; the state of bubbling agitation into which a liquid is thrown by being heated to the boiling point [followed by the suggestion that, in the 16th century, ebullition may have meant the process of extracting by boiling]" (OED). "The act of boiling up with heat. . . . [or] the commotion, struggle, fermentation, or effervescence occasioned by the mingling together any alcaline and acid liquor" (RED)

[39] *a chalybeate spring.* Chalybeate: "Impregnated or flavoured with iron, esp. as a mineral water or spring; relating to such waters or preparations" (OED).

## Page 129

[40] *Aethiop's mineral in a bolus.* Ethiops mineral: "Mercuric sulphide, HgS, prepared as a black solid by the reaction of mercury and sulphur, and formerly used medicinally as an anthelmintic [a medical agent used to expel parasitic worms, especially intestinal worms] and tonic." *bolus:* A medicine of round shape adapted for swallowing, larger than an ordinary pill (OED).

## Page 130

[41] James's fever powders. The English physician Robert James (1703-1776) patented "Dr. James Fever Powder" in the mid-1740s. It "claimed to cure fevers and various other maladies, from gout and scurvy to distemper in cattle," and was in use into the 20th century" (Welsh).

# WORKS CITED IN EXPLANATORY NOTES

Berenger, Richard. *The History and Art of Horsemanship. Volumes I and II*. London: Printed for T. Davies and T. Cadell. 1771.

Bloodgood, Lida Fleitmann, and Piero Santini. *The Horseman's Dictionary.* London: Pelham Books, 1963.

Cavendish, William. *Duke of Newcastle. A General System of Horsemanship*. 1743. London: J. A. Allen, 2000.

Chalmers, Alexander. *The General Biographical Dictionary. New Edition. Vol. VI.* London: Printed for J. Nichols et al, 1812.

Clerc, Bernard. "The Development of Equine Medicine in Europe Viewed through the Works of the Equine Veterinarians of the 17th and 18th Centuries." In Van der Horst, *Great Books on Horsemanship*, 30-37.

Courtney, W.P. "Berenger, Richard." *Oxford Dictionary of National Biography.* Oxford University Press, 2004-16.
http://www.oxforddnb.com
Accessed 2017.

de la Guérinière. François Robichon. *École de Cavalerie (School of Horsemanship)*. 1733. Trans. by Tracy Boucher. London: J. A. Allen, 1994.

Denison, Colonel George T. *A History of Cavalry from the Earliest Times. Second edition.* London: Macmillan, 1913.

Editors. "Numidia." *Encyclopedia Britannica,* 28 July 2014.
http://academic.eb.com
Accessed 2017

Editors. "Saint Augustine of Canterbury." *Encyclopedia Britannica,* 16 Dec. 2016.
http://academic.eb.com
Accessed 2018

Falkner, James. "Eliott, George Augustus." *Oxford Dictionary of National Biography.* Oxford University Press, 2004-16.
http://www.oxforddnb.com
Accessed 2017.

[RED] Fenning, Daniel. *The Royal English Dictionary*: or, *A Treasury of the English Language.* London: Printed for S. Crowder, and Co., 1761.

Freeman Strickland. *The Art of Horsemanship Altered and Abbreviated, According to the Principles of the Late Sir Sidney Medows.* London: Printed for the Author by W. Bulmer and Co., 1806.

"Hanoverian Artillery Train." Seven Years' War Project.
http://www.kronoskaf.com/syw/index.php?title=Hanoverian_Artillery_Train
Accessed 2018.

Hinde, Robert. *Discipline of the Light-Horse.* London: W. Owen, 1778.

Howell, James. *A French and English Dictionary.* London: Printed for Anthony Dolle, 1673.

Hulse, Lynn. "Cavendish, William." *Oxford Dictionary of National Biography*. Oxford University Press, 2004-16.
http://www.oxforddnb.com
Accessed 2017.

J.W., Esq. *A Military Dictionary, Explaining All Difficult Terms in Martial Discipline, Fortification, and Gunnery. Fourth Edition.* London: Printed for T. Read, [1730]. First published anonymously in 1702, the work's succeeding editions had varying attributions of authorship.

Kirby, Dr. Jeremiah. "Farriery." *Encyclopaedia Britannica. Sixth Edition* [updated from Fourth and Fifth Editions, 1810, 1817]. Edinburgh: Printed for Archibald Constable and Company, 1823, 418-573.

McDonald, Janet. *Horses in the British Army, 1750 to 1950.* Barnsley, South Yorkshire: Pen & Sword, 2017.

*Odes, Satyrs, and Epistles of Horace.* Done into English by Mr. [Thomas] Creech. Sixth Edition. London: Printed for J. and R. Tonson, 1737.

[0ED] Oxford University Press. *Oxford English Dictionary*. Oxford University Press,
http://www.oed.com
Accessed 2017.

P. Virgilii Maronis. *Georgicon. Lib. III et IV. From the Text of Forbiger, with English Explanatory Notes.* By D.B. Hickie. Verses 116-17. Cambridge: W. P. Grant. 1843.

Sidney, Samuel. *The Book of the Horse.* New Edition. London: Cassell and Company, 1893.

Tomassini, Giovanni Battista. *The Italian Tradition of Equestrian Art: A Survey of the Treatises on Horsemanship from the Renaissance and the Centuries Following*. Trans. by Tomassini. Xenophon Press, 2014.

Tyndale, W[illiam]. *Instructions for Young Dragoon Officers.* London: Printed for T. Egerton, 1796.

Urban, Sylvanus. *The Gentleman's Magazine: and Historical Chronicle. For the Year MDCCXCII. Volume LXII. Part the Second.* London: John Nichols, 1792.

Van der Horst, Koert, ed. *Great Books on Horsemanship: Bibliotheca Hippologia* Johan Dejager. Leiden: Brill, 2014.

Welsh, Jennifer. "Dr. James Fever Powder, circa 1746." *The Scientist.* October 1, 2010. http://www.the-scientist.com/?articles.view/articleNo/29277/title/Dr--James-Fever-Powder--circa-1746/
Accessed 2017.

# A TREATISE

## ON

## MILITARY EQUITATION.

### BY W. TYNDALE,

LIEUT. COL. AND MAJOR OF THE FIRST REGIMENT
OF LIFE GUARDS.

---

Fingit equum tenerâ docilem cervice magister
Ire viam, quam monstrat eques.    Hor.

——— So is my horse ———
It is a creature that I teach to fight,
To wind, to stop, to run directly on,
His corporal motion govern'd by my spirit.
            Shakesp. *Jul. Cæsar.*

---

LONDON:

PRINTED FOR THE AUTHOR,

AND SOLD BY T. EGERTON, MILITARY LIBRARY,
NEAR WHITEHALL.

MDCCXCVII.

TO

HIS ROYAL HIGHNESS

GEORGE PRINCE OF WALES,

THIS TREATISE

IS

BY PERMISSION,

RESPECTFULLY INSCRIBED,

BY

HIS ROYAL HIGHNESS's

MOST DUTIFUL,

AND DEVOTED SERVANT,

W. TYNDALE.

# LIST OF SUBSCRIBERS.

His R. H. the PRINCE of WALES[43]
His R. H. the DUKE of YORK[44]
Major General, the Earl of Harrington[45]
Right Honourable Lord Rivers, Colonel of the Dorſet Regiment of Militia.[46]
Right Honourable Lord Howard[47]

### 1ſt LIFE GUARDS.

Colonel Reed
Major Stewart
Cornet Gambier
Cornet and Sub-Lieutenant Davies

### 2d LIFE GUARDS.

General Buckley
Lieutenant George Smith

### 1st, OR KING'S REGIMENT OF DRAGOON GUARDS.

Right Honourable General Sir William Pitt, K.B.
Lieutenant Colonel Vyse, Major General
Lieutenant Colonel H. Flood
Major H. W. T. Hawley
Captain John Elliot
  John Campbell
  John Balcomb
  Robert Long
  George Teesdale
  Honourable George Irby
  T. W. Barlow
Lieutenant John Street
  W. S. Forth
  Henry Graham
  Charles Turner
  J. S. Hill
  J. Watson
Cornet John Webster
  W. Ray

### 2d DRAGOON GUARDS.

Lieutenant Adams

### 3d DRAGOON GUARDS.

General Sir W. Fawcett, K. B.
Lieutenant Colonel Payne
Lieutenant Colonel Charlton
Captain Dottin
        Davies
Cornet Harris

### 2d, OR R. N. B. DRAGOONS.

Lieutenant Colonel J. Haydock Boardman
Major Bothwell

### 7th LIGHT DRAGOONS.

Major General David Dundas
Lieutenant Colonel J. G. Le Marchant

### 10th DRAGOONS.

Colonel Cartwright
Lieutenant Colonel Slade
Major Cottin
Captain Gooch
        Fuller

Captain Seymour
      Quintin
      Honourable William Bligh
      Brummell
Lieutenant Skeen
      M'Dermot
      Chambers
      Legh
Cornet Palmer
      Wardell
      Heseltine

## 11th LIGHT DRAGOONS.

Lieutenant Colonel Carnegie
Lieutenant Colonel Childers
Major Lyon
      Gordon
Captain Parker
      Browne
      Barton
Lieutenant Mills
      Sleigh
      Bullock
      Hawker
      Horsley
Cornet Barrett
      Mabbott

Cornet Haffey
    Lutchyns
    Diggins

## 16th LIGHT DRAGOONS.

Honourable Lieutenant General Harcourt
Lieutenant Colonel Affleck
Major Lee
Captain Hawker
    Symons
Lieutenant Burnet
    Lockart
    Eminfon
Cornet Boyce
    Hay
    Bridger

## 17th LIGHT DRAGOONS.

Major Jephfon
Lieutenant Byrne
Cornet Willington

## 21ft LIGHT DRAGOONS.

Colonel Beaumont
Captain Lees

### 23d LIGHT DRAGOONS.

Lieutenant Colonel Spencer
Major Maxwell
Captain Maxwell
Lieutenant Wolseley
        Bradshaw
Cornet Robinson
        Arbuthnot

### MID LOTHIAN LIGHT DRAGOONS.

Colonel Earl of Ancram
Lieutenant Colonel Sir James Foulis, Bart.
Major Dewar
Captain Inglis
        Finlay
        Mercer

### LIGHT HORSE VOLUNTEERS

Lieutenant Colonel Herries

### FIFESHIRE CAVALRY.

Colonel Thompson

### HERTS Y. CAVALRY.

R. Whittington
Captain C. Shaw Lefevre, Efq. M. P.

### PERTHSHIRE CAVALRY.

Colonel Moray
Lieutenant Colonel Græme
Major Muir Mackenzie
Captain John Murray
   Oliphant
   Murray
   M'Laurin
Lieutenant Hay
   Moray
   Rennie
Cornet Ramfay
   Miln
   Welch
   Smart

### WINDSOR FORESTERS.

Colonel Rooke

### LATE HORSE GRENADIER GUARDS.

Major Vavasour

### LATE 120th FOOT.

Colonel Hammond

---

Reverend Mr. Salter
      Mr. E. Taylor
      Mr. William Aubrey
Mr. Moorcroft

# A TREATISE ON MILITARY EQUITATION.

## CHAPTER I.

INTRODUCTORY AND MISCELLANEOUS.

NOTWITHSTANDING an excellent work has been written by the late Lord Pembroke[48] on the art of Military Equitation, and although I by no means have the vanity to think any attempt of mine can equal that noble author's production, or that my knowledge of the art can be put in competition with his, I trust the following essay will

will not be altogether unacceptable, and flatter myself it will be found to contain some useful instructions for regimental riding-masters. These for the most part, have been taken from the ranks, and having confined their observation and study only to their profession, require explanations very clear and detailed, and the plainest and most convincing reasons when an attempt is made to alter, or introduce any thing new, as they are in general strongly bigotted to the precepts they first imbibed, however erroneous those may be.

Lord PEMBROKE has shewn a very extensive knowledge of this art, so neglected as to be almost unknown in this country, and without exception, his is the best work of the kind in our language; but I think it notwithstanding too deep and too scientific for a treatise on Military Equitation, and infinitely beyond the comprehension of those whom it is my wish to teach and reform,

form, or for the instruction of a regimental riding-master. This work shall be so plain and simple, that the commander of a regiment of cavalry, though he may never have served before but in the infantry, and not be (what every officer commanding a regiment of horse or dragoons should be) a good military-manage rider, if he approves the principle, shall have it in his power to say to his riding-master, " This, Sir, is the mode of instruction you are to adopt, and not to deviate from; and I shall, by comparing your recruits, as they are dismissed from the riding-house, with this book, be enabled to discover if you have obeyed my orders or not."

It is not my intention, nor indeed would it be my wish, were I equal to the task, to enter into a detail of the Art of Equitation; for I believe the only part of it necessary for cavalry is that which is indispensably so, to render each individual of a squadron capable

ble of executing all manœuvres required in battle. All that I wish of a horse soldier in respect to his knowledge of the art, is confined to half a dozen simple actions:

1st. To advance his horse at any pace ordered, from the halt.
2d. To turn him to either hand.
3d. To gallop to the right or left.
4th. To stop him.
5th. To rein him back.
6th. To passage to the right or left.

In all which he may be complete in three months, mounting and dismounting, and all stable duties included.

Trifling as these qualifications may appear, I firmly believe there are not in the service at this moment twenty men who possess them; and yet twenty men well acquainted with them would in three months form as many regiments of dragoons.

To instruct a dragoon or trooper[49] in an extensive knowledge of the art might be

attended

attended with many disadvantages; those which seem the most striking, are

1st. The length of time necessary for attaining this degree of perfection.

2d. The fatigue of a campaign will, in all probability, weaken the horse so far as to deprive him of his usual sensibility to the aids of his rider, and consequently make him unpleasant to one who has been accustomed to ride a lively horse; and though perhaps not so weak, or reduced, as to be quite unserviceable, yet in this state the rider would endeavour to avoid ever mounting him till he had recruited his strength and spirits; and in hopes of having him replaced by a fresh one, be tempted to get rid of him by neglect, or even lose him in the field on the first opportunity.

3d. The continual and unavoidable waving of the line would occasion false aids to be given, and worry both man and horse when too high dressed.

A school for military equitation, where regiments of cavalry might send pupils, who when sufficiently taught might be appointed *Instructors*, was an idea of Monf. le Compte DRUMMOND de MELFORD,[50] an officer whose abilities are well known, and from whose work I am not ashamed to acknowledge I extract some of the materials to complete this.

Dragoons are as useful and formidable an arm as can be brought into the field, when they are good and clever; as they are supposed to be capable of acting in both the capacities of horse and foot, but if they are not perfect, I will venture to say they are nearly useless; because a body of men who have to learn one thing only, must be more perfect than those who have to learn that, and another. Thus it is natural to suppose a battalion would be preferable to them for foot business, and that cavalry would be equally good, if not superior, to
them

them when mounted: therefore they should be perfect in both capacities; for unless they combine both, they can only be bad cavalry and indifferent infantry, or bad infantry and indifferent cavalry. The horse is however a very material part, for if he is not properly trained, instead of being an assistance and support to the man, he becomes an incumbrance. It is then absolutely necessary to teach him to be handy and manageable,

" To wind, to stop, to run directly on,
" His corporal motion governed by my spirit."

That is, the motions of the horse should be governed by the skill of the rider, who must first be instructed how to govern; for as the untaught horse is an incumbrance to a taught horseman, so will the untaught horseman be to the taught horse; thus, the untaught part of this arm, by its aukwardness, would deprive the perfect part of its power, and, in all probability, the first

affair

affair they would get into, they both would be killed.

There is a great difficulty attending the attempt to teach the manage method of riding in this country, becaufe the national one is fo very different, and however well the latter may be adapted to the common purpofes of hunting, &c. yet large connected bodies of horfemen would find it impoffible to move with that precifion neceffary, if they adopted it in their various manœuvres. The manage however, to a certain point appears well calculated to render any evolutions as practicable to numbers as to a fingle horfe. Thus it is feen in countries where the manage fyftem is unknown, that their cavalry are almoft always beaten by an inferior number of well-trained, or indeed indifferently *trained* cavalry. Notwithftanding the fuperior excellence of their horfes, the native cavalry of India were beaten by a fmaller number of Britifh dragoons

goons in the war with TIPPOO;[51] and the Hungarian huffars, unlefs immenfely overpowered by numbers, were always victorious over the Turks, though their horfes, in point of vigour, fpeed, and wind, are confeffedly fuperior to the Hungarian horfe,[52] the beft the emperor has for his cavalry. This fhews the advantage of good over bad riding, and *that* regiment whofe horfemen are individually the beft, muft neceffarily be the beft collectively. In the courfe of a campaign it will lofe fewer horfes, than any other regiment which has been worked equally with it, but which is inferior to it in point of horfemanfhip. It now remains for me to prove that the prefent fyftem, which is much the fame throughout the cavalry, is contrary to every principle of true horfemanfhip, both in the inftruction of man and horfe.

In the firft place, I fhall take notice of their method of breaking horfes. The very title given to the affiftants in the riding-houfes is enough to condemn it----Rough Riders! and very rough riders they are; gentlenefs and good temper are banifhed from the riding-houfe, and war is declared againft every animal that enters it, unlefs he is naturally more patient and meek than an afs, and even thefe qualities will not fecure him altogether from the lacerating lafhes of the mercilefs rough riders, who immediately vote him fulky. If he has a proper degree of vivacity, he is ftigmatized with the character of a *Rum Jockey*,[53] and the whip raifes a devil in him which all the fkill of the learned equeftrians cannot lay. Patience is a word not known in their vocabulary, and they expect a great unwieldy ftiff animal to be fuppled, and to acquire in ten leffons what might be looked upon as confiderable progrefs if he attained in fifty. The common effects of teaching the horfe to canter, before he has been fuppled

(which

(which they constantly do) are curbs, and not unfrequently spavins, and thus he is often spoiled, and rendered unfit for service before he quits the riding-school. The passage is never given him properly. His head is always turned the wrong way, which is the same as blinding him; for to prevent him looking the way he is going, renders him liable to run against obstructions, which if he saw he would naturally avoid. Very seldom indeed is he taught to gallop to the left as well as to the right, a neglect which leaves half the horse as stiff as ever.

Correction is often necessary, but it should be well timed and given with temper, and one blow will often suffice; for the horse recollects his fault and consequent correction; if he repeats it, the correction must be repeated; but severe flogging continued for a long time will only enrage and frighten the animal without correcting his fault.

Having thus shewn the improper method adopted by our riding-masters in horse breaking, I will point out the mistaken notion entertained of instructing the man, who unless he rides with ease and comfort to himself cannot sit for a long time on his horse; and if he is cramped and stiffened he certainly is not at his ease. See plate I. you will there, gentle reader, behold not a caricature, for it was taken from life, but a soldier placed by his riding-master, who, proud of his performance, in his own language exults over it----" Should not be ashamed for nobody to see him, if he always sat so."

In order to procure this elegant easy seat, he begins by poking up the man's chin in the air, then he squeezes his arms and his elbows close to his side, like a fowl trussed for the spit; the legs are then pulled straight down from the hips, that they hang over the saddle like a pair of tongs,

throwing

throwing him up at the same time on his fork, so that the part designed by nature to sit upon, is on this occasion relieved from its accustomed employment. This might suit one of Candide's unfortunate matrons, but is undoubtedly a very inconvenient and unsafe seat for a person who has not suffered the loss those poor ladies did.*

The various inconveniences which must arise from such a seat are obvious. That it is unnatural, therefore bad, is plain. Does the plough boy sit in that manner on his horse as he rides home from his work? No! you see him sit down plump (I make use of this word as it expresses very strongly the seat all men ought to have on horseback) his legs hang easy by his horse's side, his hand is forward holding the rein. Tell him how he may indicate to the horse, by the pressure of one leg and a turn of the

* Vide Candide.

wrist, the way he wishes him to go; teach his hand to have a *sympathetic sensation* with his horse's mouth, neither to hold himself on by the bit, nor let his horse's head hang down between his legs, and he will be excelled by very few riders in our service.

The succeeding pages will contain the method I think most proper, because it is most natural, to render a horse, of any description, tractable, pleasant and safe, and the following chapters will point out the method of riding him pleasantly and without danger. I have endeavoured to be plain and intelligible, even to those who may have never made the art their study.

CHAP.

## CHAPTER II.

OF TRAINING THE TROOP HORSE.

The great object in dressing a horse is to supple him in every joint, and every muscle, by which his limbs and his whole frame acquire a greater elasticity, and he of course becomes more active; this is not to be effected by merely lunging him, but by the hand and skill of the horseman; he will, every day he is well rode, continue to acquire more figure and ease in his gait, and become more pliable.

A horse intended only for the manage, like a good dancer, as long as he has youth, vigour, and constitution on his side, will at each lesson acquire more agility; but as it is a troop or war horse, and not a manage horse, which is the subject of this work, the method of training and pre-

paring

paring him for the squadron is what I shall endeavour to explain.

In the first place then I shall suppose he has never had a saddle on his back; it will be necessary to accustom him to undergo that operation. Great patience and much quietness must be observed by the riding-master in doing this. Having first put a snaffle bridle into his mouth, and a caveson on his nose, with the lunge buckled to the centre ring held by the riding-master; let one of his assistants approach the horse on the near side with a saddle, having a surcingle hanging over the end of it, which he must take care to hang *clear* over the horse and saddle, on either side; then place it gently on his back; this done, sooth him, and make much of him, the man continuing to hold the saddle on the horse; if he attempts to kick it off, the riding-master must raise his nose quietly, not by a snatch, as high as he can, which will prevent him.

This

This is enough for the first lesson, and should be practised only after he has finished his lesson in the lunge, as should each attempt to saddle him, until he is brought to bear it patiently, and quietly, nor should he on any account be mounted till that is the case.

The second day, having placed the saddle as before directed, let the man reach the off side of the surcingle, and bring it gently under his belly, taking care not to tickle him, and put the end of the strap through the buckle two inches, but not strain it till he has got so good a hold of it that he may be certain of securing the saddle fast to the back; not by suddenly drawing up the surcingle, but as gradually as possible; probably this will make the horse plunge a little, and should he kick the saddle off, it will be a business of time to get him to bear it at all. He should stand and feed with his saddle on, during the first lessons.

When he can bear to have the saddle girted up, and not till then, with the same gentleness put the crupper on.

## FIRST CLASS.

The first lesson to give the horse is in the lunge; and to get him to trot round to the right and left is all that can be expected the first three or four days. Each time you change him from the right to the left in the lunge, make him approach you by shortening the lunge in your hand, and caress him before you send him off again.

Having made him acquainted with the lunge, you may begin to supple him by making him bend his head and neck well into the circle, and throw out the haunches; thus making his fore legs describe a smaller circle, just within that described by the hind. Continue this lesson for ten days. After each lunging he may be mounted,
but

but must not be made to move; as all required at first is for him to suffer a man to mount and dismount. If he is good tempered, and suffers himself to be mounted without resistance, you may, towards the latter days, let him carry the man to the riding-house door, or ten yards; but by no means attempt to trot him mounted in the lunge, till the end of ten days at least.

If he has been mounted before he was recruited, on no account mount him again, till he has been well suppled in the lunge; all you can do is, if his condition will allow it, to give him longer lessons; you may gain a little time in that case, and mount him after ten days lunging instead of fourteen; that is, the four first days, comprising the first lesson, may be omitted, and you may begin at the second. This *first class* comprises only fourteen days.

## SECOND CLASS.

You may now mount him in the lunge; and having accustomed him to carry patiently in a walk, by degrees quicken him to a gentle trot; the rider bending his fore-hand into the circle, and accustoming him to the aid of the leg by pressing it against his side, until he, by flinching from it, appears to obey it, at the same time cautiously avoid giving him the spur, which should never be done but as a correction, and then decidedly. Tickling with the spur is very likely to give horses a habit of kicking at it with one leg, and some learn this trick however carefully they may have been ridden; the best mode of correcting this fault I know of, is to pinch the horse hard between both heels, and ease your hand at the same time. As soon as he has made his spring, collect him again, and try the calf of the leg to him; should he again kick,

kick, repeat the correction, till he will bear the spur---Observe, I have directed that you are to *pinch* or squeeze the horse between the heels, and not to kick him, for that very often makes a horse restive and unruly. I would not recommend a horse addicted to this fault, to be released from the lunge, till he has completely left it off.

Whilst the horse remains in the lunge, the man must make him acquainted with the aid of both hand and leg. You should raise and place his head well, never letting him get his nose down, for which reason I recommend strongly a snaffle bridle with running reins, with the ends buckled to the saddle about half way down the skirt, and to such horses as push out their noses, a head stall martingale; but to all horses a common martingale, placed so as to act on that part of the rein between the saddle and the bit. With this bridle a horse's head may be raised

very

very high and his nose well placed, if not used with violence. It is needless to add that he should be changed to the right or left frequently, and should the horse whilst in the lunge strike into a gallop, if it be not false or disunited, let him continue for a round if he pleases, but do not urge him to it. In using the *running rein*, there must be great attention shewn in the management of it. The power of it is so great, that it requires a very steady hand; in bending the horse to one hand, be careful to ease off the opposite rein: to attempt to bend the horse to the left, and at the same time to keep the right rein strained, may make him restive, as he will try every means to get rid of this irksome confinement. This is the only way to supple, and to produce an accord between the mouth, shoulders, and haunches, which if you do not, your horse is unbroke. A horse is deemed unbroke, or as a celebrated French writer says, "il n'est pas fait, s'il

n'est

n'est pas uni des epaules, de la main & des jarrets."[55]

The horse may now be released from the lunge and worked round the house. First in a slow well collected trot, then in a faster trot, and lastly in his full trot; which two last paces must not be attempted till he performs the slow pace with ease and exactness, keeping time as it were to music. In the course of this second class, which will contain twelve lessons, he may be brought to his fastest trot, if he shews a disposition to be active, for it is obvious, that all horses are not equally stiff or equally supple, some requiring to be kept for a much longer time at the different exercises recommended in each class, but none for less.

## THIRD CLASS.

This third class will contain eighteen days or lessons, and much must be done in this time: he must be put upon his haunches and

and finished all but galloping, which is the last thing taught, and the easiest if the horse has been properly worked in the foregoing classes.

The beginning of each day's lesson should be trotting, with frequent changing, and on small circles, and should finish with what is generally called *shoulder in,* (but which in fact is not, that being one of the most difficult airs attempted in the manage;) To proceed in going to the right, bend his fore-hand by shortening the right rein; I say his fore-hand, because the legs and chest must be *half faced* to the right, and preserve that position as he advances, his hind legs moving straight, and carrying the greatest part of the weight on them; to make them do which, hold your horse in, on the mouth-piece of his bridle, and throw your heels back, threatening him with the spur: do not hold him too hard, or he will rear up; but if he is properly held, he will be found to champ on his bit, and yield with his under jaw to its preſſure. As this leſſon is very ſevere, it ſhould be ſhort and frequently repeated: for example, the horſe ſhould not be conſtrained after *once* he *has obeyed*, for more than one long ſide, and one corner of the houſe; releaſe him then by degrees, and immediately repeat it. The paſſage muſt alſo in each day's leſſon of this claſs be given thus; put him in the ſame attitude required for the ſhoulder-in, but inſtead of forcing him up to his bit, preſs him hard by degrees with the calf of the oppoſite leg, and he will move ſideways; take care however he does not go back, (which horſes in learning this are apt to do) and that his fore-hand always leads. Reining back is neceſſary to be taught in this claſs, and care ſhould be taken to make him move back in a ſtraight line; horſes in the firſt attempt at this are very aukward, and throw their *croups* from ſide to ſide.

*The bit* is never to be put in the mouth till he has been galloped and finished.

A very neceffary accomplifhment which is never practifed, fhould alfo, and during the remaining time he is detained in the fchool, be taught both to man and horfe; I mean leaping; not flying, but cool ftanding leaps; if the horfe during the remaining claffes was, until he could perform tolerably well, firft led over, and then rode over a bar at the finifh of each leffon, he would with fuch practice leap any hedge or other low fence, which now proves an infurmountable obftruction to moft of the light and to all the heavy cavalry.[56] When in the drill, after his difmiffal from the fchool, he might be practifed at ditches.

The fucceeding clafs which is the fourth, contains eighteen leffons or days, and is intended for the gallop.

FOURTH

## FOURTH CLASS.

OF THE GALLOP.

In the first attempt at this pace, choose that hand to which the horse is most supple; for as men are more active and perfect with their right hand, generally speaking, than with their left, from being accustomed to use the right hand most, so horses naturally acquire a greater agility with one leg than with another, from using it most when colts; and I have remarked, that either from this cause (or what I think is more likely, from the person who broke him having ridden stiffer to one side than to the other) there is a greater suppleness in one side than in the other. After the cavefon and lunge are fixed on his head, order the rider to present him to the gallop, which is done by putting your horse together, that is, setting him on his haunches, and feeling both reins equally, and applying both legs to his sides, but the outside harder than the inside. This will throw his croup into the circle, and oblige him to strike off with the proper leg, which when he has done, sustain him, or keep him to it by threatening with your whip, assisting with your legs and voice, but not using your spurs. When he gallops tolerably well to the right, (which requires six lessons at least) he must be taught to gallop to the left, which will also require six lessons more, and six succeeding ones out of the lunge will be necessary to teach him to change on the gallop from right to left. The bit may now be placed in his mouth, and for the space of six days more repeat every lesson of the third and fourth classes, that his mouth may be accustomed to the bit by the steady hands of the instructors previous to his being delivered to the men for the drill.

The

The foregoing courſe has taken up ſixty-two days, or, Sundays included, ten weeks, which with perſons properly qualified to act as horſe-breakers, is as ſhort a time as can be, conſidering the kind of horſe on which our light and heavy cavalry are mounted; but is enough. I then recommend a ſufficient number of the beſt horſemen in the regiment to be ſought for, and a drill to be formed of all horſes in the ſame ſtate of forwardneſs, to be commanded by one of the rough riders when leaping the bar and the ditch. The horſe evolutions muſt be practiſed; the two firſt days in a walk, the ſix ſucceeding ones in a trot, and four more in a gallop. I ſhould now conſider the horſe fit for the ſquadron when properly accoutred.

The ſuperiority a regiment riding well and on well broke horſes would have over one where the horſe required as much room and as much time to turn as a broad wheeled waggon,

waggon, is too obvious to need any further comment.

I must however beg leave to notice a curious fact, on which I will afterwards hazard an opinion. The Hussars are, generally speaking, men of five feet nine inches high, and they ride horses little more than fourteen hands. One would imagine a man so much *under horsed* would be useless rather than formidable; but the astonishing agility of these people, and the service they have performed, sufficiently shew the contrary. I am therefore led to imagine the reasons they are so under mounted, are---first, that unless a Hussar be extremely active, and can turn almost as quick as a man on foot, he is useless; therefore agility is required in the animal more than strength; but that is no reason why both agility and strength should not be combined in him if possible. Secondly, a horse with agility and strength, equal to

the

the weight of a man five feet nine inches high, is both expensive and scarce; and as a Hussar ought to be easily and cheaply replaced, there generally being a great expenditure of them, as they are always at work, the preference was given to the activity of the animal, and the Hussar was put upon a small horse of low price; but like all other small horses, of much more agility than a dragoon horse. From his small size, also were derived one other advantage, and, in my opinion, one great disadvantage. The advantage is, that the little horse fills himself sooner, and takes less forage to do it than the large one; and the disadvantage is, the Hussar cannot leap, being too low and too heavily loaded to attempt it. How capitally might a regiment of Hussars be mounted in Great Britain and Ireland, where there are abundance of horses able to carry from twelve to fourteen stone, about fifteen hands high,

bred

bred on the commons, particularly in Wales, to be purchased for twenty pounds a piece. They are not very handsome, but they have shoulders and haunches, capable of very active movements. A regiment thus mounted, the men from five feet three inches to five feet six, would be superior to Hussars, either for speed, agility or strength, nor would they by any means form a despicable regiment of dragoons, I mean in point of weight.

Thus concludes my system of horse-breaking,[58] taking up seventy-two days, Sundays not included, not quite three months; and as for the most part recruit horses seldom are more than four years old, they should on no account be worked harder; six, seven or eight years old horses, may by taking double lessons, if they are in good condition, be finished in half that time.

CHAP.

## CHAPTER III.

OF THE INSTRUCTION OF THE RECRUIT ON HORSEBACK.

I shall preface this chapter by obferving, that although this art may be reduced to a fcience, I think it unneceffary to inftruct the recruit in the theory of its principles. But as this is meant for the ufe of thofe who are to become inftructors, it is indifpenfible to explain the theory of it.

I have made the foregoing obfervation left they might be led into the error of conceiving, that detailed explanations were neceffary for the recruit. I would tell him that when he wants to turn his horfe to the right, he muft place the bridle hand in the fifth pofition, and apply his right leg to the horfe's fide.

It here becomes neceſſary to ſpeak more particularly on the poſition of the hand.---The left is called the bridle hand, from its always holding the reins. The arm ſhould fall naturally and with eaſe from the ſhoulder to the elbow, which ſhould be a little advanced, ſo that you may juſt perceive the light between the elbow and the body. The wriſt ſhould be in a line with the elbow, and the hand over the pummel of the ſaddle; yet this poſition of the wriſt can only be laid down as a general rule; for much muſt depend on the manner in which the horſe carries his head; if high, the wriſt muſt be dropped a little to lower his noſe; and if low, muſt be held higher to raiſe it. But when the wriſt is once placed, the elbow ſhould be immoveable, as all motion ſhould proceed from the wriſt alone.

The hand is now to be placed in the firſt poſition, viz. the nails ſhould be oppoſite the buttons of the waiſtcoat, and of courſe
the

knuckles towards the horse's mane, the little finger being on a line with the bone of the elbow. From this position it is capable of performing four movements, or taking four other positions for various purposes, which will hereafter be explained, and some of which are accompanied by a help or movement of one or both legs. The hand is generally carried in the first position, when the horse is going forward, and not shewing any disposition to turn on either side, to throw up his head, or to make any false movement.

The second position consists in turning the knuckles upwards, and then slackening the rein as after halting; or to give the horse the liberty of moving forward by easing his mouth, and this is usually accompanied with closing the legs to the horse's side.

The third position is to check the horse, in order to make him drop from a quick pace into a slower one, or to stop him entirely,

tirely; and requires the knuckles to be turned downwards. No greater or wider movement of the hand can be neceffary, or ought to be allowed, either to flacken the rein, even in the fafteft pace, or from that pace to ftop the horfe, than what is called the fecond and third pofitions.

The fourth pofition confifts in carrying the little finger to the left, and inclining the knuckles upwards, by which movement the horfe will turn his fore-hand to the left, and if the left leg be applied to his flank at the fame time, he will turn as long as the hand and leg are continued in thefe pofitions.

The fifth pofition confifts in carrying the little finger outward, to the right, and inclining the knuckles downwards, and if the right leg be preffed to the flank, the horfe will do the fame to the right as he did to the left in the fourth pofition.

FIRST

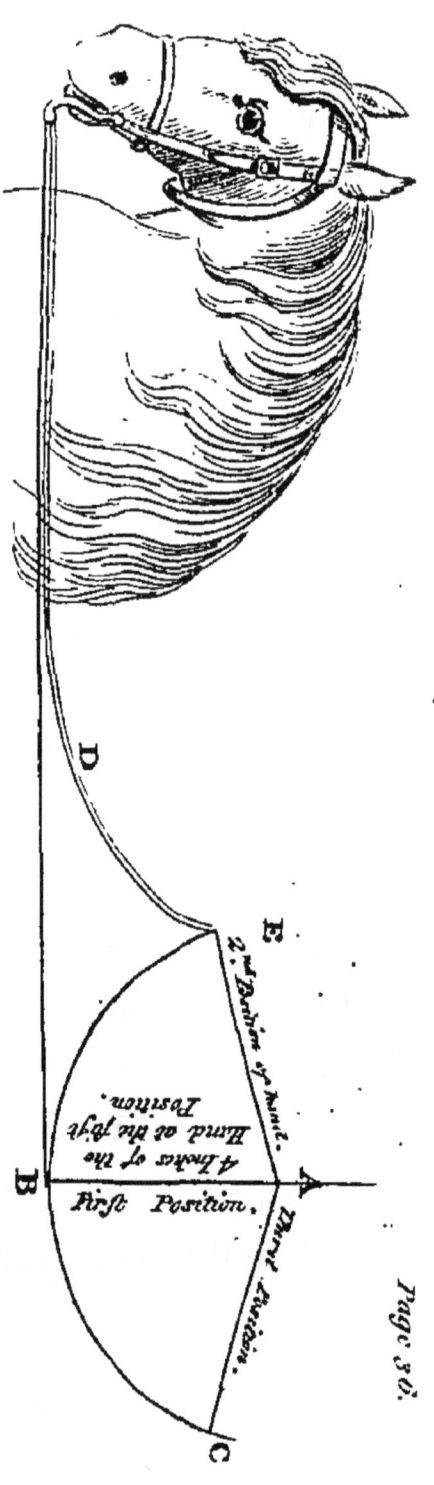

D. Shews that by moving the toe B.A. into the Position E.A. That the Rain must be Slack, as it shortens the distance from the Bit to B, very near 4 Inches, The hand is Supposed to be 4 Inches from the Top of the Knuckles of the first Finger to the lower side of the little Finger.

## FIRST CLASS.

I now proceed to divide my inftructions into claffes, in the fame manner as was done in the preceding chapter.

I fuppofe the recruit to be fo far advanced in his foot-drill as to be quite fuppled; if he comes to the riding houfe before he has attained that, he will probably make but flow progrefs. He fhould firft be taught to faddle his horfe in fuch a way that the faddle may not pinch or bruife the withers, and this muft be firmly impreffed in his mind, by being repeated to him every day during his remaining in the firft clafs. The faddle fhould be placed about fix or feven inches behind the withers, and the crupper buckled fo as to prevent it, whilft ungirted, from getting forwards; but when the girts are once drawn, the crupper fhould not be on the ftretch, for that would not only fet the

the horse a kicking, but gall his tail, the consequence of which often renders a horse unserviceable for several weeks. It may produce a violent inflammation and sore, not to be cured but with great time and attention. To avoid this, you must instruct your recruit never to have his crupper tight, and that if the saddle will not keep in its place without the crupper, it does not fit the horse properly, and the recruit should be instructed to apply to his quarter-master to get it altered. You next tell him to be very careful to draw the buckle of the girth two or three inches at least on both sides above the bottom of the pannel, or else the buckles will fret and gall the skin. If the pannel be thin and hard, he should report it to his quarter-master, and desire to have it new stuffed, or a new one. Make him unsaddle and saddle his horse in your presence, or that of your assistants, and see that he pays attention to every particular in detail. The
riding-

riding-mafter is certainly the perfon who has, or ought to have bitted the horfes of the regiment; I therefore fuppofe the horfes to have the bridle properly placed in the mouth.

Of the fhape and properties of the bit, a moft powerful machine when properly conftructed and managed, but ineffectual and likely to do mifchief when improperly conftructed or managed, I fhall fpeak very fully in another place.

The recruit fhould be taught that the bridle is to hang in the horfe's mouth, fo as not to pucker up the corner of it, but juft to touch it, and the mouth-piece will then reft on the bars, which is the proper place for it: that the curb chain muft be placed neat and fmooth, and ought never to be loofer, than to admit the finger between it and the jaw. The nofe band buckled juft in the fame degree of tightnefs round the nofe over the bridoon head-ftall, which

muft

must be precisely of the same length with the bridle.

You then proceed to give him the method of mounting and dismounting, which in the heavy dragoons is made much longer than in the light, by an useless set of motions. I think the following motions would be as good as any.

At the word "*Prepare to mount,*" let the man who is standing on the near side of his horse's head, turn quickly to the right, stepping back at the same time, so as to be placed opposite the shoulders; then let him take up the reins as quick as possible, and putting the little finger of the left hand between the near and off rein, let him grasp both together with the three other fingers, along with a lock of the mane, and shorten the reins a little, by drawing the ends with the right hand through the left, and thus letting them fall on the off shoulder.

The right hand then will be occupied by holding the ſtirrup: at the next word, "*Mount*," let the man put his left foot in the ſtirrup, and at the ſame moment his right hand to the back part or cantle of the ſaddle; let him ſpring nimbly off the ground with the right foot, aſſiſting himſelf with his hand, and throwing his whole weight on the left foot in the ſtirrup. Then he ſhould raiſe his right thigh as high as he can, to clear the cloak, or whatever may be on the pillion or pad, wheel himſelf round into his ſeat, and all this ſhould be done in leſs than a ſecond.---Being there ſeated, he ſhould let go the mane, and with his right hand draw the bridoon rein behind the left, and place it in it, which he immediately raiſes to the firſt poſition, at the ſame time placing his right foot in the ſtirrup. I ſhould have remarked that the firſt thing to be done on the recruit's leading

Page 41.

his horse into the riding-school is to put on a caveson and lunge.

The man when mounted may be confidered as consisting of three parts---two moveable, and one immoveable. The body from the head to the hips, and the legs from the knees downwards, are moveable. This leaves the thighs immoveable. In placing the man, cautiously avoid touching or even mentioning his feet or legs; if you do, or attempt to place the leg (as it is called by the riding-masters) you will certainly give a stiffness to it, which will for ever prevent his becoming a horseman. Begin with his head, and tell him to hold it back, not *up*; for holding up the head is throwing up the chin; but holding *back* the head is drawing it up from resting on the collar-bone, and placing it even between the shoulders. Let his arms hang down on each side, as if they were coat-sleeves stuffed. By bending

the

the small of the back, and forcing the lower part of the chest forwards, the shoulders will be carried rather behind than in a direct perpendicular.

## OF THE THIGH AND SEAT.

The common method used by riding-masters to give a man a seat, is to place him exactly on the fork, neglecting to bring that part designed by nature for the seat, on the saddle, and then seizing him by the foot, to twist his legs as a dentist does a tooth to loosen it. This stiffens the knee joint, and gives the legs and thighs the same position in the saddle that a pair of tongs would have; whereas if they would content themselves with telling the man to turn *the outside of his thigh* well up, that is, to force it round towards the horse's head, and not to stiffen the knees, the consequence would be, that the leg would turn with it, and come into the proper place.

Having placed the man in his saddle, give him his reins in his left hand, properly divided, as directed in the mounting instruction, (page 40,) and put a switch in his right, directing him to carry it in the manner the regiment carry their swords. This will be the means of accustoming him to ride afterwards with more ease when he has his sword given him.

A young beginner should not be hurried; before he is made to trot in the lunge, he should feel himself a little confident, and it ought to be from him, or at his desire, that the horse should quicken his pace; his being without stirrups, if you begin too roughly with him, will, for the first lessons, make him either very unsteady, or else he will cramp and stiffen himself to keep on his saddle, by holding on with his legs.

legs. He should be lunged fourteen days at least, first on easy horses, very slowly, and as you perceive him acquire a seat, quicken the trot by degrees.

He should practise frequent changes to the right and left, and at each change observe that he puts his hand to the proper position. When you see that the movement of the hand begins to be a little familiar to him, he may be instructed in the helps with the legs; which are as follow :---First, to indicate to your horse your intention of moving forwards or faster.

Secondly, to make him place his croup in any situation you please, to make him passage, and to turn him to either side.

The legs and spurs are also made use of as corrections. In giving the leg, take care not to use *the spur*, unless the horse will not obey you; therefore begin with fixing the knee well to the saddle, and then turn out the toe, apply very gently the calf of
the

the leg to the horse's flank, carrying the foot *well back*. If touching the side with the leg gently does not produce obedience in the horse, squeeze him harder; if he will not answer that, pinch him with the spurs; and the next time you threaten him with your leg, you may depend on his obeying your movement---but avoid, if possible, using the spur. Thus when the legs and hand have a sympathetic sensation, assisting each other mechanically, the horseman may be told, that his hand and legs are in unison, which is the best translation I can give of the French term, "*l'accordance de la main & des jambes*,"[58] without which a man cannot be a horseman.

I have said the legs must assist mechanically the help of the hand; by this I mean, that the horseman should contract such a habit of using his legs to assist the hand, that though he may be occupied by talking to company he is riding with, yet if the road

has

has a sudden turn to the right, the hand should mechanically fall into the fifth position, and the right leg be at the same instant applied as above directed, to his flank, to turn his croup, without discontinuing the conversation.

I have before observed, that some of the helps of the hand are accompanied with aid from the one or both legs. Thus, at the second position, which is to ease the hand, or rather the mouth by releasing it from the pressure of the bit; if, as may be the case to a dull horse, this motion does not sufficiently indicate to him your intention, apply both legs gently to him, and he will immediately move forward; but if one only is applied, it will make him only move his hind quarters *from* the pressure; this shews therefore clearly, that if at the same time the hand is put in the fourth or fifth position, which will inevitably turn his forehand, the leg be applied, it will hasten the evolution by

sustaining

sustaining his hind-quarters. For, if you bend a horse's forehand only, and then put him in action, he will move on a circle, instead of turning on a pivot, which the pressure of the leg obliges him to do.

To return to the pupil, (who as he is only to be taught mechanically) must be accustomed by practice to acquire your method of riding; therefore observe that at each turn or changing of hands, *his* legs and hands act together. If at the end of fourteen days trotting and practice he has acquired the habit of using his hands and legs at each turn, he may be released, and begin the second class.

This, as well as the first, should be gone through without stirrups, which should not be given till he is dismissed from the school. During the first three weeks or eighteen days, nothing but trotting, with frequent changes and circles at each corner, is necessary. The next eighteen lessons or days,
after

after a few rounds of trotting and changing, shew him how to passage, and to rein back, and finish each lesson by a few wheelings.

The third and last division of this class is twelve days, and should be employed in forming up from single horsemen following one another to two, four, six, and eight, and then reducing this again in the same way; also varying it. Thus from single horsemen forming a front of any given telling or number, such as *Form Front* of eight; and for which reason, when the recruits arrive at this state of forwardness, they must be told off into files, threes, and sections of eight. Wheeling by threes, which is the method at present adopted for a squadron to break up in, to go about, or for gaining ground towards a flank, should be practised; and trifling as this simple evolution appears, it certainly is difficult, for few squadrons ever do it well. It cannot be done too slowly, as by wheeling on the center horfe, which muft turn on its own ground, one horfe muft wheel forwards round, which is eafily done, but the other flanker muft wheel backwards *round*, which is very difficult. He muft alfo in this divifion of the fecond clafs, be inftructed how to turn to the right or left, and to the right or left about. I recommend the leaping bar at a low hole at firft, for the finifh of each day's leffon, as he quits the fchool, on a horfe that can leap tolerably well.

OBSERVATIONS ON THE FOREGOING CLASS.

The time I have allotted for the exercifes of this fecond clafs is very fhort; but the impatience of commanding officers is very often fo great, that they will fcarcely allow time for any thing to be taught the recruit, but mounting and difmounting, before they put him into fquadron.

In the firft clafs of this chapter the recruit

cruit learns very little but to get a tolerably steady seat on his saddle, his horse merely carrying him; the second class perfects his seat, and teaches him to ride, that is,

To move forward from the halt in  
To stop from  
To turn to the right or left in  
} a walk, trot, or gallop.

To retain the horse from  
To quicken him from  
} one pace to another.

To rein him back and to passage.

With respect to the first of these objects: The hand eased, and the legs applied to the sides, will make him walk off; the same motion, with a harder squeeze of the leg, will make him trot off.

The gallop is explained in the third class of this chapter.

Of the stop or halt. Whatever degree of retention you would indicate to your horse, the motion of the hand will always be the third position; for it is by an imperceptible sway of the body backward, that the horse is stopped in all paces; however from the full trot and the gallop, in order to stop the horse on his haunches, which should always be done, the bridle hand should also be gently and gradually raised, at the same moment the body inclines backward, and the legs likewise, whenever the full stop is marked to the horse, should be near his side, and the heels rather back; this will oblige him to sustain the whole weight of himself and his rider on his hind quarters.

This explains the use of the second and third position of the hand, and how far the legs can assist; it also shews the method of reining back, which is a greater degree of retention than the *full stop*, and requires more sway of the body, and more elevation of the hand, but must not be continued; therefore immediately on the horse's shewing

ing his obedience by going backwards, turn your hand to the firſt poſition.

If you intend going farther back, replace your hand and body in the attitude which firſt produced the motion, and ſo continue eaſing and retaining alternately till you have completed your intention. The legs ſhould be near the horſe during the whole action to be ready to aſſiſt in the direction of the croup.

To turn to the right, or right about, the hand is put in the fifth poſition, and the right leg ſupports the croup.

In the turn to the left, or left about, the hand is in the fourth poſition, and the left leg ſupports the croup. It is to be obſerved that a horſe ſhould never turn on his fore legs, if it can poſſibly be avoided, but there ſhould be deſcribed a circle round the hind legs; however, turning on the fore legs is in ſome inſtances unavoidable, as in the

center

center horse of threes, when endeavouring to turn on his own ground.

The application of the leg to the croup is to prevent it; horses very frequently lame themselves by turning on their fore legs, by causing a violent extension of some muscle of the leg, or by throwing a greater weight on that limb, than it ever was intended to carry. The passage is explained in the third class of the preceding chapter.

## THIRD CLASS.

The gallop is the next pace to be taught, nor can there, in my opinion, be any other instruction for the man, than that which an instructor imparts by his hand and leg to the young horse; I shall therefore refer my reader to the fourth class of the chapter on Training the Troop Horse, as any instruction I can imagine for the man, would be only a repetition of what I said in that place,

place, except that a less time will be necessary in the lunge, and two days practice to each hand; however it will require a much longer time out of the lunge, as it will be necessary for the man to practise a great deal in filings, formings, and circles, as directed in the third division of the second class. This being finished, the recruit should be dismissed from the school, and a drill formed and practised as directed for compleating the recruit horse.

The sword exercise, an excellent and necessary accomplishment, should now be practised. On this exercise I will make one observation, which is, that the men who have been generally selected by regiments to learn this art, have not been good horsemen, and have therefore introduced false principles of riding, such as leaning forwards, abandoning the horse, and urging him forward, without the least command

mand over him, to the very great danger of the rider. This, however, I am told, is by no means either adopted or approved of by the very deserving officer to whom the army is indebted for this additional and useful exercise*.

I must now warn my reader of two bad habits which dragoons are very apt to contract: The first is, wide movements with the hands, partly owing to their never holding their reins short enough.

The second, that of always throwing their body forward on the expectation of the word to *march*, previous to their putting their horses in motion, in the same manner, as badly drilled men, when waiting for the word, will rise on tiptoe.

Nothing can be worse than this, and very

---

* Lord Heathfield, whose skill in equitation amply repays the great pains he bestows upon it, told me, that Major Le Marchant disclaimed any such erroneous practices.

great care should be taken to correct this fault, and almost all recruits have a natural tendency to it. I was obliged once in the case of a man, who was so much addicted to it, as not to be conscious of doing it, to tie a string from his tail to the crupper; the check this gave him when he attempted to rise, reminded him of his fault, and a few lessons perfectly corrected him.

It is highly proper and a most excellent practice to make a ride halt frequently, and move off together in a trot; this should be performed in the third class of this chapter, as also changing at the *wall* by a right or left about turn. In short after you have got over the first division of the second class; the more you vary the lesson the better.

Throughout these two courses of equitation in the instruction both of man and horse, I have given to each a specific number of days; I do not mean that it should be inferred from hence that no *more* time

I should

ſhould be employed in each claſs, but I am convinced that leſs will not do; however, that muſt be left to the diſcretion of the inſtructor.

I have endeavoured to methodize it by giving the courſe a regular ſucceſſion of exerciſes, increaſing the difficulty of each according to the progreſs of the learner.

## CHAPTER IV.

EVERY field-day, and every time a dragoon mounts his horse, attended by his officers, he should be made to observe a strict attention to good riding, and great care should be taken by every officer commissioned or non-commissioned, (for which reason serjeants and corporals should have a little more instruction than privates) to correct any fault or bad riding in the course of exercise; and if it appeared to arise from ignorance or inattention, to report such observation to the riding-master, who, it is almost needless to add, should attend, but not in the squadron on these days. One day in the week, all the year round, should be set apart for the whole regiment, with the officers, to go to a riding drill; a plan of which will be given; however before I begin upon that subject, I cannot help observing how few and trifling the instructions are, which most regiments give a young officer, who though he may have led the field with fox-hounds, and rode over turnpike gates in his native county, is in all probability, totally unacquainted with the method of riding which is necessarily adopted in regiments of cavalry, and which must be observed by every individual, for one bad rider will cause such confusion as totally to prevent a squadron from moving with regularity and precision.

The method I have recommended, so far from spoiling a hunting seat, will, in my opinion, give it more force, and be more favourable to the horses; it will also make them go up to, and take their leaps in a much better stile.

## RIDING DRILL FOR A REGIMENT, TROOP, OR SQUADRON.

When on ground sufficiently spacious, form a column of troops with intervals of about five times the length of the front of a troop. Each troop should be provided with at least six strong stakes pointed with iron, and an iron collar round the top to prevent the stake from splitting, when driving into the ground. The troop should then be told off into files from the right, and into six sections. Opposite the nose of the pivot horse, set up one stake, and call it No. 1; an officer then casting his eye along the line from the wheeling flank horse's nose must *aline*, on the stake No. 1, a quartermaster, or any man not of the front rank. This done, order your front rank *only* to file straight off by the stake, No. 1. When the rank is in file, at twenty yards beyond

the

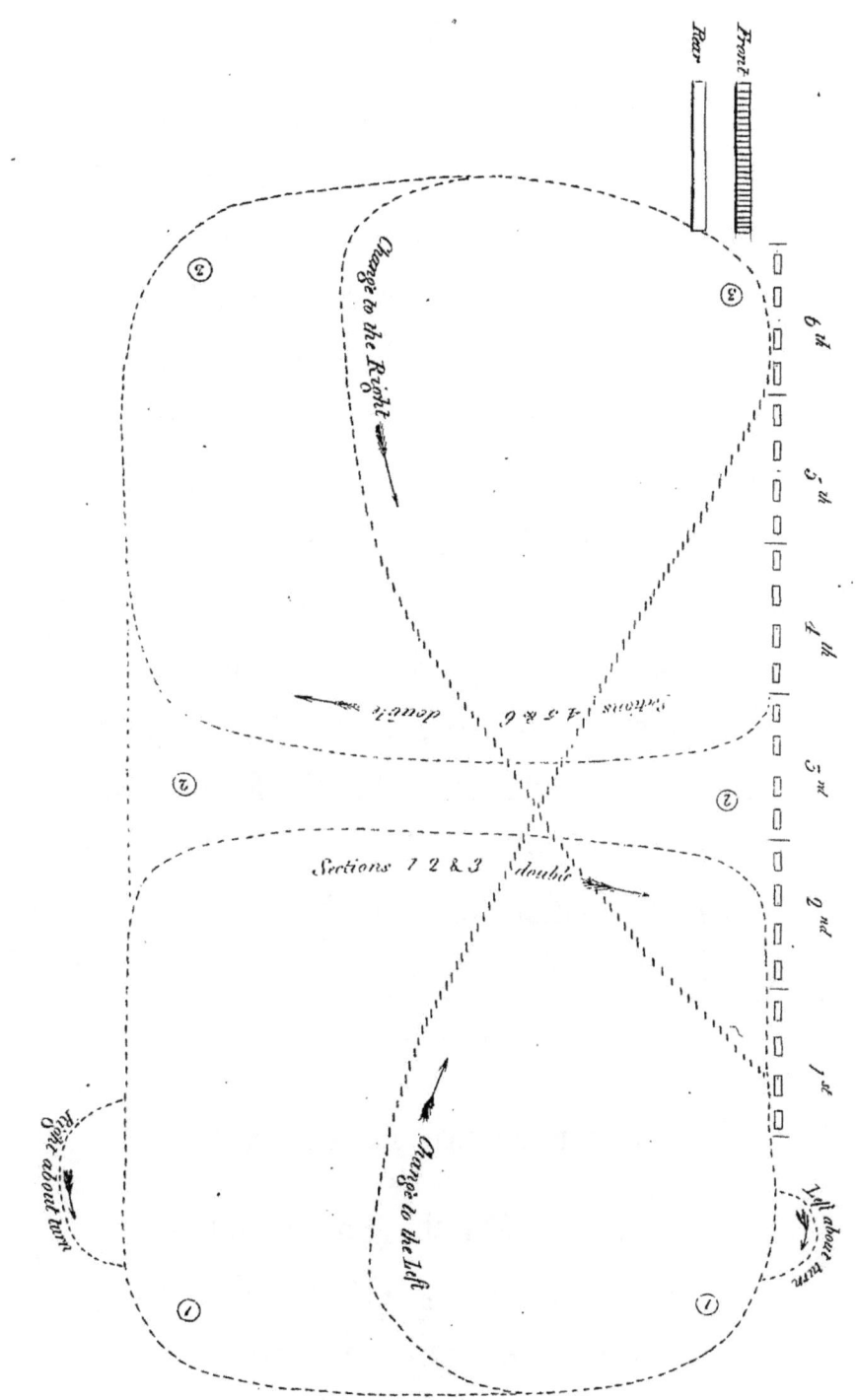

the leading horse, set up another stake, and that will be the length of the drill ground. Opposite the center put another stake; call these 2 and 3; the width of the ground should be about one third of the length; this may be easily guessed, and thence on a line parallel to the first, set up the other posts marked No. 1, 2, 3; so that No. 1 at each corner may be opposite No. 3. Each troop should do the same. The rough rider and officers mounted should be within the square, and give the lesson as follows, which consists of three divisions or reprisals.

## LESSON.

#### FIRST REPRISAL TO THE RIGHT.

Trot. March (one round)
*Sections* 1, 2, *and* 3.---*Double from* 2 *to* 2. That will make half your ride; cut the ground in the middle, and then when the other

other half which are sections 4, 5, and 6, come opposite the next number 2;

*Sections* 4, 5, *and* 6.---*Double.*

When this is done, CHANGE.

Doublings to the left to be repeated in the same manner as before described.

CHANGE.
{
Halt. Walk. March.
Trot. Walk. Halt.
Left about turn.
Trot. March.
Right about turn. } Without
Left about turn. } halting.
Halt. File to the rear.
}

Rear rank advance.

*(Repeat the first reprisals.)*

N. B. Whilst the rear ranks were unemployed at the ride, the sword exercise or any other might be practised which would not fatigue the horses; the front might do the same during the interval of each reprisal.

## SECOND REPRISAL TO THE RIGHT.

Trot. March. Form Section.

1, 2, and 3, double.

4, 5, and 6, double.

CHANGE $\begin{cases} \text{Halt. Walk.} \\ \text{March. Halt.} \\ \text{Trot. March.} \end{cases}$ Each section moves at the same moment.

*Reduce the sections.*

(This brings them to single horse-men again.)

This is a left movement on going to the left. $\begin{cases} \text{CHANGE.} \\ \textit{Form sections.} \\ \text{(Doublings as to the right.)} \\ \text{Halt. Walk. March.} \end{cases}$

After the rear rank has gone through this reprisal, begin the third in a gallop well supported, horses well on their haunches.

## THIRD REPRISAL.

To the right.---Gallop. March.

In order to do this well and with safety, let the leading file be opposite the left stake (No. 3.) the others in his rear, three good horses lengths asunder, as each man must on the word march, and at the same moment, make his horse strike into a gallop with the proper leg.

(Twice round.)

Form sections.

   1, 2, and 3, double.
   4, 5, and 6, double.

CHANGE! (Twice round.)

Form sections.

 1, 2, and 3, ⎱
 4, 5, and 6, ⎰ Double.

Halt. Gallop. March.
Trot. Walk. Gallop.
Halt. Rein back. March. Halt.

    K     March.

March.

Halt.

N. B. The word march *only*, always signifies Walk!

To the right, } Paſſage. Halt.
To the left,

The field officers and riding-maſter ſhould not attach themſelves to any particular troop, but keep moving from one to the other, aſſiſting in the inſtruction. In the riding houſe this drill would occaſionally be very uſeful to officers and quartermaſters.

An excellent exerciſe for a regiment or ſquadron to practiſe, as it makes the men ride well, and at the ſame time accuſtoms them to preſerve a uniformity of pace, conſiſts in putting the whole regiment into a column of ſixes (which a wheel of ranks by three's to the right or left, will do) with a ſpace of half a horſe between each diviſion. Then ordering it to move off at the word march,

march, in any pace, the whole should start at the same time, and halt all together. This is very difficult: a regiment however, perfect in this exercise, would find all others very easy.

## OF STIRRUPS.

In the chapter on the instruction of the man I have said nothing of stirrups, which ought not to be considered as giving any strength or force to a horseman's seat, tho' by a too frequent practice of riding with them, men very often make them a material part of it, which gives a stiffness to the whole body, and renders it impossible to ply to the motions of the horse, and from this circumstance the rider becomes more liable to be thrown off. The intention of the stirrups is to relieve the thighs and hips from the weight of the legs, and to raise and support the foot.

The usual mode of ascertaining the proper length of the stirrup consists in bringing the tread of the stirrup to the ancle bone; this however shortens the leather too much. The method I prefer, if the man cannot be relied on to adjust the stirrup himself, is after he is well seated on his saddle, to raise the foot at the ball with your hand, till the man bears on it without raising the heel; then with your whip measure up to the bar on the saddle-tree, from whence the stirrup hangs, and shorten the leather till the tread is the same length from the bar as what you marked on your whip.

I have not spoken of accustoming horses to fire or the sword, because time and patience, with gentle usage, are the only means to make the animal familiar with them. It is however safest, and consequently adviseable on all occasions, when a horse is to be initiated in any new exercise, which may astonish his sight or hearing, to secure
him

him with a cavefon and lunge, and attempt it, if poffible, in the beginning without a rider. With thefe implements there is power fufficient to hold him, and length of rein to give him room to fpring without his entirely efcaping, which would not be the cafe with the bridle reins. It is not uncommon to fee a frightened horfe, either by the force of his jerk, or by fighting with his fore legs, get away from a man; and when once he has found out that it is in his power fo to do, he will be very apt to repeat it.

## OF THE BRIDLE.

Riding-mafters are very fond of making the men ride on both bit and bridoon, and this they affert tends to preferve the horfe's mouth, than which, in my opinion, nothing can be more erroneous; I have met with others who give better reafons for this practice,

practice, viz. that the men might raise their horses by the bridoon. However trifling this may appear, there is nothing which requires so much judgment as the manner of using the bridoon to raise the head, and the bit to place it; for it is only for this purpose both can be made to act at the same time, and then it must be done very steadily and carefully, and requires a much greater skill in the art than a dragoon either has, or can stand in need of. For the first reason, viz. that of preserving the horse's mouth, to any one who considers and understands the effect of the bit (which all riding-masters ought to do) it is clear that the bars of the mouth, having, as is the case in a common coach bit, a uniform pull on them, must soon lose their sensibility and become callous. A well broke horse is accustomed to carry his head high, and by feeling the bit lightly on his bars, preserves in all paces this attitude. The fear of the bit acting

acting on them, prevents his hanging his head on it; but if the bridoon is strained, the horse bears upon it to a certain point, and consequently requires a stronger power of the bit to make him feel; and much greater indeed than would be necessary did nothing press the bars: In time therefore he loses that dread of the bit, and throws himself on it, as he did on the bridoon; this renders him hardmouthed. A heavy and insensible hand which never yields to the horse, also produces the same effect: A heavy hand is a fault, not easily corrected; however, the few and confined movements of the hand, I think necessary to work the bridle, and have consequently recommended in the foregoing part of this book, remedy this defect in a great degree, the men not having it in their power to annoy or spoil their horses mouths.

Another bad effect of making men ride with the bridoon is, they acquire such a habit

bit of making ufe of both hands, that when they have their fwords drawn they cannot ride at all, for want of the bridoon. Their horfes are uneafy, owing to the unfteadinefs of the bridle hand which throws their nofes up in the air. Bridoons are a very neceffary appendage to the bit, not to affift its powers, but to relieve the mouth occafionally. Thus on marches over level and good ground, and in going to and from exercife, they fhould always be ufed, but then alone and as a fnaffle bit, letting the reins of the bit hang on the horfe's neck. The hands when ufing this bit muft be low, or high, according as the horfe carries his head, but the men fhould not, as is the cafe almoft always at watering parades, pull up their horfes nofes, but endeavour to place the head in the fame pofition the bit would bring it to.

## OF THE BIT.

Bits in general are extremely ill conſtructed, owing to an erroneous opinion that the longer the branch is, the more likely it is to ſtop a horſe. This will appear clearly to be a miſtaken notion, when the attitude in which a horſe is ſtopped by a bridle is conſidered. It muſt be the ſame, though in a leſs degree, as he is in when reining back, which is indiſputably carrying his weight on his haunches. The effect of a long branch is to a certain point the ſame as a properly conſtructed bit, that is, it pinches the under jaw on the bars; but from its length and power, it pulls the whole of the head down, and the noſe into the cheſt, which of courſe prevents the horſe from throwing himſelf on his haunches, as it brings the weight on his ſhoulders; and no horſe on his ſhoulders can be ſtopped

ped firm and short. This, a bit well made, obliges him to do; as by squeezing the bar between the mouth piece and the curb chain, the horse is obliged to yield his under jaw, which he cannot well do without raising his head a little, and if at the same time the horseman assist this motion of the horse by raising his hand, and throwing his body gradually back, the horse is absolutely obliged to halt on his hind legs, and thus in my opinion, a bit with a branch just clearing the chin is generally sufficient, but an inch longer is quite enough in any case. The mouth piece should not be small, for if it is, it cuts and lacerates the bars, and renders the mouth uncertain. The chain should for the same reason be made very thick. A bit should be as pleasant as possible to the horse; if it hurts and annoys him, it takes off his attention to the slight movements of the hand, and makes him fretful and unpleasant. No horse can

run

Fig. 2. A. Holster fixd & made a part of the Saddle.
B. Necessary Bag made a part of the Saddle.
C. Opening of the Necessary Bags.

Page 75.

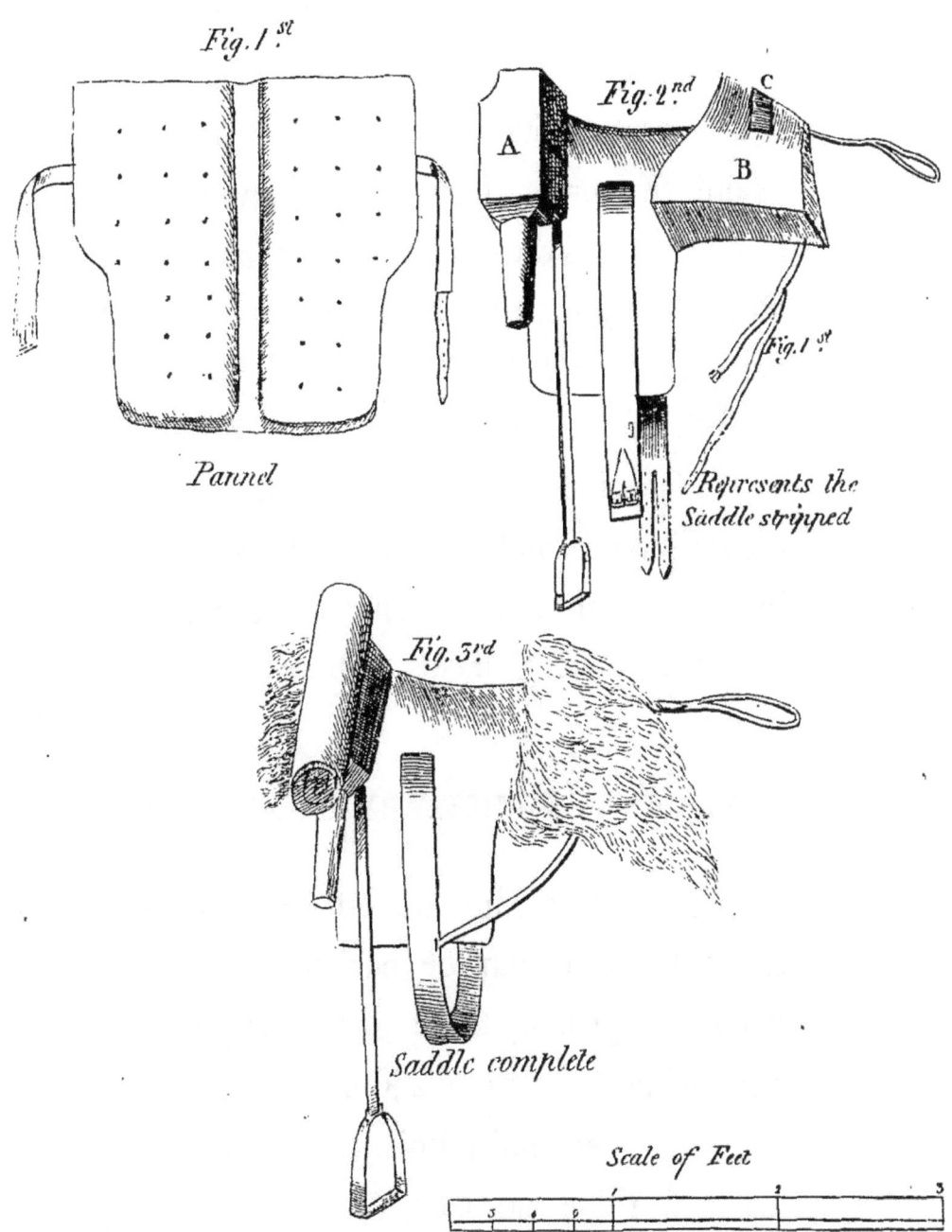

Fig. 1.st

Pannel

Fig. 2.nd

Represents the Saddle stripped

Fig. 3.rd

Saddle complete

Scale of Feet

run away if his head is kept up, and the bit acts on the bars; but if the bit cuts and torments him, he will throw his nose up, and in that position, the bit, by falling against the corner of the mouth, cannot act on his bars, and then he becomes unmanageable. The mouth piece should not be longer than the width of the horse's mouth, just suffering the branches to hang down close by the side of the lips. The hollow of the mouth piece should not be so wide as to allow the corner to press upon the bars, which would very much torment and hurt the horse.

PLAN FOR A MILITARY SADDLE.

A soldier's saddle should be so constructed that it may be put on the horse without displacing the luggage, which is not trifling, at every unsaddling.

Holsters and pistols.
Carbine and bucket.

Cloak; and

Necessary bag,

In which should be carried his clothes, his stable utensils, oil, and all other things requisite for cleaning his horse, arms, and appointments.---Such articles, which in time of peace might be called necessaries, but on service would be dispensed with, might be carried in a valise or cloth, upon the necessary bag; which, if it could be made a part of the saddle, would be preferable, as straps and cords are liable to be broken; and not securing the luggage firm, causes warbles and sore backs. This, I think, might be done without much increase of weight, and being covered with leather or painted cloth, be rather ornamental, and at the same time keep the necessaries dry. I have attempted by a drawing to explain how such a saddle might be made.

FIGURE 1. This is a pannel, which is made to tie into the saddle as a saddle cloth does.

does. It has a surcingle on it, which secures it tight to the back, and prevents it hurting the back. It is also easier dried than if it was nailed to the saddle, which is a very important consideration.

Figure 2. The holsters form two burs, as to a French pique saddle; it rises very much at the pummel, to help the rider's seat, and is then perfectly straight; but at about fourteen inches from the bottom of the slope of the pummel, a strong piece of leather four inches high, is nailed on, resembling the hind cantle of a pique saddle. At about six inches from this nail is another piece of leather, which will be at the extremity of the saddle; this back bit should be only two inches high. The saddle tree which carries this, will continue out under it, about as wide as a pad or mail pillion, and under this the necessary bag should hang, the sides of which are formed by the two pieces of strong leather before-mentioned.

tioned. A quilted pannel, such as smugglers use to keep the load off the horse's flank, and a stout piece of leather sewed over this and joined to the two sides, compose the bottom of the bag; the upper part and ends are formed of one piece of supple leather, sewed to the upper edges of the side, and it opens immediately in the center by a slit on the top. The stirrups hang from the bottom of the swell of the holster pipe, which as a surcingle is sewed to the bottom of the skirts, is more convenient than hanging from the tree; it also leaves the legs more at liberty. The holsters are made stiff and strong, and staples fixed to them for the cloak to be buckled to. Over the pummel on the inner edge, a ring is put for the center cloak-strap, which keeps it off the pistols; a flounce hangs over the holsters, under the cloak. The housing is already described: all the other appointments of bucket, breast-plate, &c. may be easily added. A saddle of this kind,

kind, which is not much longer than a common one, nor much, if at all, heavier, would be put on the horse in at least one-sixth of the time it now takes up.

I make no doubt but some objections may be made to this saddle, which is of my own invention, though the idea is borrowed from the old pique, which I have frequently heard experienced people call a good sort of military saddle, as being most easy, and giving the man great support in his seat; and the objection that it heats the men, is answered, by observing how much the cloak, necessaries, and perhaps three day's forage, must on any sort of saddle do it. It is an undenied fact, that soldiers, when fatigued, sleep on their horses, and the load of forage they carry, from its forming a support for their back, is an additional inducement to them to give way to their desire to sleep. On the ordinary saddle they sit loose and roll about, which more frequently produces

warbles

warbles and fore backs than any other cause; but in the pique saddle the thighs are steadied by the burs and cantle, which circumstance gives a great security and strength to the seat, and keeps the man firm in his saddle even whilst asleep. I should not have obtruded this saddle of mine on the public, but for the perpetual complaints which are made respecting that imporant appointment. However, from the recent circumstance of a German officer\* of rank being sent to England, to procure patterns of our saddles, which are to be adopted by the army he serves in, I am inclined to think them preferable to any now used.

\* This was written in February, 1797.

*FINIS.*

# EXPLANATORY NOTES TO A TREATISE ON MILITARY EQUITATION

## Title page

[42] Fingit equum tenera docilem cervice magister
Ire viam, quam monstrat eques.                               Hor.

The Jocky trains the young and tender Horse,
While yet soft mouth'd he breeds him to the Course.
(Horace, *Odes, Satyrs, and Epistles*, 276)

## Page vii

[43] *His R.H. the PRINCE of WALES.* (1762-1830). Baptized under the names of George Augustus Frederick, the eldest son of the fifteen children of George III was created Prince of Wales in 1762 and ascended to the throne as George IV in 1820 (Hibbert, *ODNB*).

[44] *His R.H. the DUKE of YORK.* (1763-1827). Prince Frederick, second son of George III, was gazetted a colonel in the army in 1780; appointed colonel of the 2nd Horse Grenadier Guards in March, 1782; promoted major-general in November, 1782; and made lieutenant-general in 1784, when he became colonel of the 2nd or Coldstream Guards. He was created duke of York and Albany in 1784 (Stephens, *ODNB*).

[45] *Major General, the Earl of Harrington.* (1753-1829). Charles Stanhope, styled Lord Petersham in 1756, and created third earl of Harrinton in in 1779, was an army officer who advanced from ensign in the Coldstream Guards in 1769 to colonel of the 1st Life Guards in 1792, major-general in 1793, and lieutenant-general in 1798. Made a general

in 1803, he served as commander-in-chief in Ireland from 1805 through 1812. "A highly respected professional soldier," he introduced a sword that "became the standard army sword in the early nineteenth century" (Farrell, *ODNB*). Tyndale dedicated his manual, *Instructions for Young Dragoon Officers* (1796) to Harrington (Tyndale was major of the First Regiment of Life Guards and Harrington major-general).

[46] *Right Honourable Lord Rivers, Colonel of the Dorset Regiment of Militia.* George Pitt, first Baron Rivers (1721-1803), MP, politician, and diplomat, assumed a seat in the House of Commons in 1742, and, upon his creation as Baron Rivers of Stratfieldsaye, Southampton, in 1776, in the House of Lords. Pitt was appointed colonel of the Dorset militia on its establishment in 1757 (Barker, *ODNB*). Pitt was succeeded in 1799 by Richard Bingham (1740-1823), formerly captain in the militia and father of Sir George Ridout Bingham (1777-1833), a distinguished officer in the Penisular War (Anon. 396-97).

[47] *Right Honourable Lord Howard.* Presumably Kenneth Alexander Howard, first earl of Effington (1767-1845), the only child of Captain Henry Howard and his second wife, Maria. Howard became an ensign in the Coldstream Guards in 1786; was promoted lieutenant and captain in 1793; captain-lieutenant and lieutenant-colonel in 1797; and brigade major in 1798. Following several commissions in the Peninsular War and elsewhere, he became a lieutenant-general in 1819 and a full general in 1837 (Barker, *ODNB*).

## Page 1

[48] *Lord Pembroke.* Henry Herbert, 10th earl of Pembroke (1734-1794), author of *Military Equitation: A Method of Breaking Horses, and Teaching Soldiers to Ride* (1761-93).

## Page 4

[49] *dragoon or trooper.* Dragoons were mounted soldiers who could fight from horseback or on foot. "In late 16th century Europe, a mounted soldier who fought as a light cavalryman on attack and as a dismounted infantryman on defense. . . . From the early wars of Frederick II the Great of Prussia in the 18th century, dragoon has referred to medium cavalry. The light cavalry of the British army in the 18th and early 19th centuries was for the most part called light dragoons" (Hosch, "Dragoons"). A military dictionary from the early 18th century offers a more colorful definition: "Musketeers mounted, who serve sometimes a Foot, and sometimes a Horseback, being always ready upon any Thing that requires Expedition, as being able to keep Pace with the Horse, and do the Service of

Foot. In Battle, or upon Attacks, they are commonly the *Enfans Perdus,* or Forlorn, being the first that fall on. In the Field they encamp either at the Head of the Army, or on the Wings, to cover the others, and be the first at their Arms" (*A Military Dictionary*).

## Page 6

[50] *Mons. le Compte Drummond de Melford.* Louis-Hector Drummond de Melfort (1721-1788) spent most of his life in the French cavalry, eventually rising to the rank of Lieutenant-Général in 1780. Drummond wrote a treatise on cavalry tactics, *Essay sur les évolutions de la cavalerie* (1748) that circulated widely in manuscript for nearly three decades and eventually became the basis for his *Traité sur la cavalerie* (1776), "one of the most influential books on military equitation of the 18th century" (Van der Horst, 608). In *Instructions for Young Dragoon Officers* (1796), Tyndale invokes Drummond de Melford as "a most able tactician" and summarizes Drummond's "plan for the distribution of an advanced guard, showing also how to search the country through which the column has to pass" (Tyndale, *Instructions*, 75-79).

## Page 9

[51] *the war with Tippoo.* Tippu (or Tippoo) Sultan (1750-1799), Tiger of Mysore, fought successfully against the Hindi Marathas in the 1760s and 1770s. He also defeated the British Colonel John Brathwaite in 1782 to end the second Mysore War. Having concluded peace with the British in 1784, however, he subsequently provoked a British invasion in 1789 that led, after two years of warfare, to his signing the Treaty of Seringapatam, in Britain's favor, in 1792. Tippoo was killed in battle in a fourth Mysore War launched by the British in 1799 (Pletcher). For contemporary views on the last two conflicts, see Major Dirom, *A Narrative of the Campaign in India, which Terminated the War with Tippoo Sultan, in 1792* (1793), and R.K. Porter, ed., *Narrative Sketches of the Conquest of the Mysore, Effected by the British Troops and Their Allies, May 4, 1799* (1800).

[52] *Hungarian hussars.* "Member[s] of a European light-cavalry unit employed for scouting, modeled on the 15th century Hungarian light-horse corps. The typical uniform of the Hungarian hussar was brilliantly coloured and was imitated in other European armies. . . . Several hussar regiments of the British army were converted from light dragoons in the 19th century" (Hosch, "Hussar"). A military dictionary from the early 18th century again offers a more colorful definition: "Horsemen, cloathed in Tygers and other Skins,

and garnished and set out with Plumes of Feathers; their Arms are the Carbine, Pistols and Saber. . . . [When attacking, they] fall with such Vivacity on every Side, that, unless the Enemy is accustomed to them, it is very difficult for Troops to preserve their Order. When a Retreat is necessary, their Horses have so much Fire, and are so indefatigable, their Equipage so light, and themselves such excellent Horsemen, that no other Cavalry can pretend to follow them; they leap over Ditches and swim over Rivers, with a surprising Facility" (*A Military Dictionary*).

## Page 10

[53] *Rum Jockey.* "Rum" is a British colloquialism, meaning, in this pejorative context, "bad, spurious, suspect"(*OED*), thus, a malicious or untrustworthy horse.

## Page 13

[54] *Candide's unfortunate matrons [with footnote: Vide Candide].* Tyndale is alluding to Voltaire's *Candide ou l'Optimisme* (1759), a widely popular and raucous picaresque work published in English as *Candide, or, All for the Best* (1761). Tyndale obviously is making a joke about genitalia, though the exact point escapes me. Candide's "unfortunate matrons" include Cunégonde (the woman he loves), The Old Woman, and Paquette, all of whom suffer sexual abuse either through rape or prostitution.

## Pages 22-23

[55] *. . . as a celebrated French writer says, "il n'est pas fait, s'il n'est uni des epaules, de la main & des jarrets."* Tyndale writes in the preceding sentence, "This is the only way to supple, and to produce an accord between the mouth, shoulders, and haunches, which if you do not, your horse is unbroke." Tyndale might be referring to Claude Bourgelat, invoked by Pembroke (see Explanatory Note to Pembroke, page 61). Chapter VII of Berenger's translation of Bourgelat's *Nouveau Newcastle* (1754) begins: "The End which the Horseman proposes to attain by his Art, is to give to the Horses, which he undertakes, the *Union*, without which, no Horse can be said to be perfectly drest" (Berenger, 54).

## Page 26

[56] *Light and heavy cavalry.* See Introduction, note 6.

**Page 29**

[57] *Horse evolutions.* The term *evolution*, in general usage, refers to "a movement or change of position," and, in its specific military usage, to "a manoeuvre executed by troops or ships to adopt a different tactical formation" (*OED*). See, for directly relevant examples, Tyndale, "Horse Evolutions," in *Instructions for Young Dragoon Officers*, 38-39; Hinde, "Horse-Evolutions for the Light-Troop" and "Evolutions and Attacks of Ditto," in *The Discipline of the Light-Horse*, 74-89; Baron von Steuben, "Evolutions of the Horse," in *The Manual Exercise, and Evolutions of the Cavalry*, 89-95; or *The Light-Horse Drill: Describing the Several Evolutions in a Progressive Series* (1800), in its entirety (the last work designed specifically for the "volunteer corps of great Britain").

**Page 32**

[58] *Horse-breaking.* "Breaking" has been the conventional English term for the initial training of a horse, including acceptance of bridle, saddle, and rider, since the first treatise in English on dressage and equitation, Thomas Blundeville's *The Arte of Ryding and Breakinge Greate Horses* (1560). Though "breaking" or training methods became increasingly less harsh over the following three centuries, the term itself appears not to have acquired a negative connotation until the early 20th century, when we find Piero Santini, for example, saying "there never was such a word as 'breaking' in [Caprilli's] vocabulary" (Santini, 18; and see Williams, v).

**Page 46**

[59] *l'accordance de la main & des jambes.* Tyndale translates this himself a few lines earlier as "hands and legs in unison." Pembroke writes, similarly, that "the legs should not only be corresponding with the hand, but also subservient to it" (Pembroke, 55).

**Page 56**

[60] *the very deserving officer [with footnote to Lord Heathfield].* General George Augustus Elliott was created Baron Heathfield in 1787. See Explanatory Note to Pembroke, *Dedication*.

**Page 77**

[61] *pique saddle.* See Explanatory Note to Pembroke, page 17.

**Page 80**

[62] *a German officer of rank being sent to England, to procure patterns of our saddles [with a footnote indicating February, 1797].* Discussing riding saddles of the regular British army in *Horses and Saddlery* (1965), Major G. Tylden writes that Frederick, Duke of York, traveled to Germany in 1780 to study the Prussian Army and returned in 1787 and 1791. During the latter trip, at the request of the Prince of Wales, he sent home samples of saddles for both heavy and light cavalries. In 1796, the British cavalry adopted a new pattern saddle, based on the Prussian cuirassier (heavy) saddle. Tylden reports that "a German Officer was sent over to procure patterns [for possible use in the Prussian cavalry]." Tylden cites Tyndale as his source, however, so provides neither clarification nor corroboration of Tyndale's reference (Tylden, 127-28).

# WORKS CITED IN EXPLANATORY NOTES

Anon. "Bingham, Richard." *The Annual Biography and Obituary, for the Year 1825* [and 1824]. Volume IX. London: Longman, Hurst, et al, 1825.

Barker, G. F. R., *rev.* S. Kinross. *Oxford Dictionary of National Biography*. Oxford University Press, 2004-16.
http://www.oxforddnb.com
Accessed 2017.

Baron de Steuben. Regulations for the *Order and Discipline of the Troops of the United States* [and] *The Manual Exercise, and Evolutions of the Cavalry.* New York: Greenleaf's Press, 1794.

Berenger, Richard. *A New System of Horsemanship: From the French of Monsieur Bourgelat.* London: Printed by Henry Woodfall, for Paul Valliant, 1754.

Farrell, S. M. "Stanhope, Charles." *Oxford Dictionary of National Biography*. Oxford University Press, 2004-16.
http://www.oxforddnb.com
Accessed 2017.

Herbert, Henry. 10th Earl of Pembroke. *Military Equitation*. Fourth edition. London: Printed for G. and T. Wilkie, 1793.

Hibbert, Christopher. "George IV." *Oxford Dictionary of National Biography*. Oxford University Press, 2004-16.
http://www.oxforddnb.com
Accessed 2017.

Hinde, Robert. *Discipline of the Light-Horse*. London: W. Owen, 1778.

Hosch, William L. "Dragoons." *Encyclopedia Britannica*.
http://academic.eb.com
Accessed 2017.

Hosch, William L. "Hussar." *Encyclopedia Britannica*.
http://academic.eb.com
Accessed 2017.

*Odes, Satyrs, and Epistles of Horace*. Done into English by Mr. [Thomas] Creech. Sixth Edition. London: Printed for J. and R. Tonson, 1737.

[*OED*] Oxford University Press. *Oxford English Dictionary*. Oxford University Press
http://www.oed.com
Accessed 2017.

Pletcher, Kenneth. "Tippu Sultan." *Encyclopædia Britannica*, 26 Aug. 2014.
http://academic.eb.com.
Accessed 2017.

Santini, Piero. *The Forward Impulse*. New York: Huntington, 1936.

Stephens, H. M., *rev.* John Van der Kiste. "Frederick, Prince." *Oxford Dictionary of National Biography*. Oxford University Press, 2004-16.
http://www.oxforddnb.com.
Accessed 2017.

Tylden, Major G. *Horses and Saddlery: An Account of the Animals used by the British and Commonwealth Armies from the Seventeenth Century to the Present Day with a Description of their Equipment*. London: J. A. Allen, 1965, 1980.

Tyndale, William. *Instructions for Young Dragoon Officers*. London: Printed for T. Egerton, 1796.

Van der Horst, Koert, ed. *Great Books on Horsemanship: Bibliotheca Hippologia Johan Dejager.* Leiden: Brill, 2014.

Williams, Richard F. "Publisher's Introduction." *Principles of Dressage and Equitation*, by James Fillis. Franktown, VA: Xenophon Press, 2017.

# XENOPHON PRESS LIBRARY

www.XenophonPress.com

Xenophon Press is dedicated to the preservation of classical equestrian literature. We bring both new and old works to English-speaking riders.

*30 Years with Master Nuno Oliveira,* Henriquet 2011

*A New Method to Dress Horses,* Cavendish 2018

*A Rider's Survival from Tyranny,* de Kunffy 2012

*Another Horsemanship,* Racinet 1994

*Austrian Art of Riding,* Poscharnigg 2015

*Classic Show Jumping: the de Nemethy Method,* de Nemethy 2016

*Divide and Conquer Book 1,* Lemaire de Ruffieu 2016

*Divide and Conquer Book 2,* Lemaire de Ruffieu 2017

*Dressage for the 21st Century,* Belasik 2001

*Dressage in the French Tradition,* Diogo de Bragança 2011

*Dressage Principles and Techniques,* Tavora 2017

*Dressage Principles Illuminated, Expanded Edition,* de Kunffy 2017

*École de Cavalerie Part II,* Robichon de la Guérinière 1992, 2015

*Equine Osteopathy: What the Horses Have Told Me,* Giniaux 2014

*Fragments from the writings of Max Ritter von Weyrother,* Fane 2017

*François Baucher: The Man and His Method,* Baucher/Nelson 2013

*Great Horsewomen of the 19th Century in the Circus,* Nelson 2015

*Gymnastic Exercises for Horses Volume II,* Eleanor Russell 2013

*H. Dv. 12 German Cavalry Manual of Horsemanship,* Reinhold 2014

*Handbook of Jumping Essentials,* Lemaire de Ruffieu 2015

*Handbook of Riding Essentials,* Lemaire de Ruffieu 2015

*Healing Hands,* Giniaux, DVM 1998

*Horse Training: Outdoors and High School,* Beudant 2014

*I, Siglavy,* Asay 2018

*Learning to Ride,* Santini 2016

*Legacy of Master Nuno Oliveira,* Millham 2013

*Lessons in Lightness,* Mark Russell 2016

*Methodical Dressage of the Riding Horse,* Faverot de Kerbrech 2010

*Military Equitation: or, A Method of Breaking Horses, and Teaching Soldiers to Ride,* Pembroke, and *A Treatise on Military Equitation,* Tyndale, edited by Charles Caramello, 2018

*Principles of Dressage and Equitation, a.k.a. Breaking and Riding,* Fillis 2017

*Racinet Explains Baucher,* Racinet 1997

*Science and Art of Riding in Lightness,* Stodulka 2015

*The Art of Riding a Horse or Description of Modern Manège in Its Perfection,* D'Eisenberg 2015

*The Art of Traditional Dressage, Volume I DVD,* de Kunffy 2013

*The Ethics and Passions of Dressage Expanded Edition,* de Kunffy 2013

*The Forward Impulse,* Santini 2016

*The Gymnasium of the Horse,* Steinbrecht 2011

*The Horses, a novel,* Elaine Walker 2015

*The Italian Tradition of Equestrian Art,* Tomassini 2014

*The Maneige Royal,* de Pluvinel 2010, 2015

*The Portuguese School of Equestrian Art,* de Oliveira/da Costa 2012

*The Spanish Riding School & Piaffe and Passage,* Decarpentry 2013

*To Amaze the People with Pleasure and Delight,* Walker 2015

*Total Horsemanship,* Racinet 1999

*Training with Master Nuno Oliveira double DVD set,* Eleanor Russell 2016

*Truth in the Teaching of Master Nuno Oliveira,* Eleanor Russell 2015

*Wisdom of Master Nuno Oliveira,* de Coux 2012

Available at www.XenophonPress.com

www.ingramcontent.com/pod-product-compliance
Lightning Source LLC
Chambersburg PA
CBHW060504240426
43661CB00007B/913